建筑立场系列丛书 No.

家居生态
Ecologies of Domesticity

汉英对照
（韩语版第375期）

韩国C3出版公社 | 编

杜丹 于风军 孙探春 徐雨晨 | 译

大连理工大学出版社

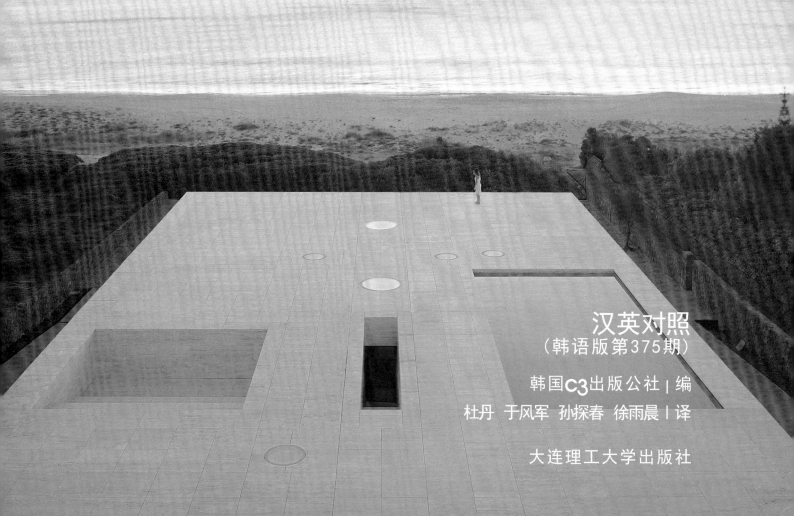

作为荒废环境中的新型产业的艺术

004 作为新型产业的艺术 _Jack Self

010 布达艺术中心 _51N4E

020 Arquipélago当代艺术中心
_Menos é Mais Arquitectos + João Mendes Ribeiro Arquitecto

036 iD Town满京华美术馆 _O-office Architects

048 普拉达基金会艺术博物馆 _OMA

068 卡托维兹西里西亚博物馆 _Riegler Riewe Architekten

082 敦刻尔克地区当代艺术基金会 _Lacaton & Vassal

住宅设计
家居生态：
景观设计的三种方式

096 家居生态：景观设计的三种方式 _Nelson Mota

102 永恒之屋 _Alberto Campo Baeza

116 蒙桑双宅 _João Paulo Loureiro

132 卡特彼勒住宅 _Sebastián Irarrázaval Arquitectos

146 八岳山住宅 _Kidosaki Architects Studio

152 MO住宅 _Gonzalo Mardones Viviani

160 "秋之屋" _Fougeron Architecture

170 向日葵住宅 _Cadaval & Solà-Morales

182 桧原住宅 _SUPPOSE DESIGN OFFICE

190 建筑师索引

建筑立场系列丛书 No.59

Art as the New Industry
at the Abandoned

004 *Art as the New Industry _ Jack Self*

010 Buda Art Center _ 51N4E

020 Arquipélago Contemporary Arts Center

　　 _ Menos é Mais Arquitectos + João Mendes Ribeiro Arquitecto

036 MJH Gallery of iD Town _ O-office Architects

048 Prada Foundation Art Museum _ OMA

068 Silesian Museum in Katowice _ Riegler Riewe Architekten

082 FRAC Dunkerque _ Lacaton & Vassal

Dwell How
Ecologies of Domesticity
Three Ways of Designing the Landscape

096 *Ecologies of Domesticity: Three Ways of Designing the Landscape _ Nelson Mota*

102 House of the Infinite _ Alberto Campo Baeza

116 Two Houses in Monção _ João Paulo Loureiro

132 Caterpillar House _ Sebastián Irarrázaval Arquitectos

146 House in Yatsugatake _ Kidosaki Architects Studio

152 MO House _ Gonzalo Mardones Viviani

160 Fall House _ Fougeron Architecture

170 Sunflower House _ Cadaval & Solà-Morales

182 House in Hibaru _ SUPPOSE DESIGN OFFICE

190 Index

作为荒废环境中的新型产业的艺术

　　西方国家普遍接受的一个说法是，限制工业化意味着工厂的关闭，把产品外包给东方国家，大片市区（有时候是整个城市）被废弃。对此，西方经济体希望逐渐用无形的商品来促进经济增长，金融服务、管理培训和咨询这些在促进经济增长方面都是至关重要的，但是更多时候，城市取得全球性成功的关键因素是创意产业与文化产业。这种重新定位表明了产品生产地点的巨大转变，然而，人们却不去想在所谓的东方国家正在发生什么样的变化。实际上，东方国家科技的快速发展和巨大的社会变革正在形成与西方国家相似的条件。在深圳，废弃的工厂并不意味着本土制造的结束，而是意味着把它迁至效率更高、技术上更加先进或是更加现代化的工厂。与此同时，中国的文化产业也在迅速增长：现在年轻的一代不再像父辈那样成为体力劳动者，而是正在国际舞台上寻求艺术认同感。全球化导致了大批建筑的荒废，其中不乏许多质量相当好的建筑。然而，如今的全球化也催生了人们对艺术创作和提高文化影响力的空间的需求，前工业建筑因此获得了第二次生命。

The accepted narrative in the West is that deindustrialisation means the closure of factories, the outsourcing of production to the East, and the ruination or abandonment of large urban territories (and sometimes whole cities themselves). In response, Western economies increasingly hope to use immaterial production to propel growth. Financial services, management training and consultancy are all vital, however it is more often creative and cultural production that is seen as a key measure of a city's global success. This recount describes a huge shift in the location of production, however it rarely considers what might be happening in the so-called East, where rapid technological development and colossal social change are in fact creating similar conditions. In Shenzhen, abandoned factories do not signify the end of local manufacturing, but rather signify its relocation to more efficient, technically sophisticated or modern plants. At the same time, Chinese cultural production is exploding: younger generations that would previously have become labourers are now seeking artistic recognition on the international stage. The forces of globalisation have resulted in huge numbers of useless buildings — many with truly remarkable qualities. Now those same forces are creating a need for spaces of artistic production and cultural influence, and ex-industrial architecture is getting a second life.

Art as the New In
at the Abandoned

布达艺术中心/51N4E
Arquipélago当代艺术中心
/Menos é Mais Arquitectos + João Mendes Ribeiro Arquitecto
iD Town满京华美术馆/O-office Architects
普拉达基金会艺术博物馆/OMA
卡托维兹西里西亚博物馆/Riegler Riewe Architekten
敦刻尔克地区当代艺术基金会/Lacaton & Vassal

作为新型产业的艺术/Jack Self

Buda Art Center/51N4E
Arquipélago Contemporary Arts Center
/Menos é Mais Arquitectos + João Mendes Ribeiro Arquitecto
MJH Gallery of iD Town/O-office Architects
Prada Foundation Art Museum/OMA
Silesian Museum in Katowice/Riegler Riewe Architekten
FRAC Dunkerque/Lacaton & Vassal

Art as the New Industry/Jack Self

当我们谈起这些被改造成现代文化场馆（艺术画廊、艺术家工作室、展览馆、影剧院等）的前工业建筑的时候，大部分时候我们只是在描述建筑物本身。这些研究所缺少的正是这些作品的背景：这些工业建筑为什么会被遗弃？它们真的是衰退或者年代久远的标志吗？为什么要改造？为谁而改造，用哪里的资金来改造？

资本主义的核心原则之一就是工人必须以尽可能快的速度出卖自己的劳动力。当他们打工的这个地方待遇无法与对手公司或产业竞争时，他们就必须换地方以赚取更多的工资。为了保证生产中的供需平衡员工流动势在必行；如果（打个比方来说）相对于钢材的成本来说，钢铁工人的数量过多，那么工人的工资就会下降；工资降了，一些工人就会换工作，那么供需平衡也就恢复了。

在全球范围内，无论是在一个国家内还是国与国之间，也是同理；如果城市比农村发达，可以预见的是，人口会单一流向比较发达、更加有利可图的地区。当然，从一个更复杂的层面来讲，劳动力的流动其实就是企业、产业、城市、地区和国家彼此竞争的结果。要想强大，它们必须具有吸引力，不仅仅要吸引劳动力，还要吸引资本投资。

本文所提到的六个建筑作品，每一个都是废弃的工业建筑得到巧妙的改建和重新设计的例子，每一个作品背后都有不同的故事。把它们相互关联的因素就是当地社区在成功改建和设计这些建筑中所起到的重要作用。这种地方色彩与这些建筑在国际经济竞争中的作用直接相关，其目的或在于开拓新兴市场，或是为了增强疲软的市场。

When we talk about ex-industrial buildings that are repurposed into modern cultural venues (art galleries, artist studios, exhibition spaces, theatres and so on), we mostly just describe the architecture. What is missing from these studies is truly the context of the works: why were the industrial structures abandoned? Are they really signs of decay, or simply age? Why have they been converted, for whom, and with what money?

One of the core principles of capitalism is that workers must always try to sell their labor at the highest possible rate. When their place of work fails to compete with a rival company or industry, they must be prepared to travel in order to make more money. This imperative towards mobility is supposed to guarantee supply and demand in production: if there are too many steel workers in relation to the cost of steel (for example) then wages will fall, some workers will change jobs and the balance will be restored.

At a global scale, this is also true both within and between countries; and where cities outperform rural areas there is a predictable one-way population flow toward more profitable territories. Of course, one complex dimension to worker mobility is that companies, industries, cities, regions and nations are all competing with each other. To grow they must be attractive, not just to labor, but also to capital investment.

The six works of architecture featured in this article, each one an abandoned industrial complex that has been skilfully restored and repurposed, all have very different stories. What unites them is how much local communities have at stake in their success. This localism is directly tied to their role in global economic competition, whether that means creating new markets or reinforcing weakening ones.

The Arquipélago Contemporary Arts Center by Menos é Mais Arquitectos with João Mendes Ribeiro Arquitecto is located on the island of São Miguel in the Azores archipelago – islands frequently referred to as the best kept secret in Europe. Re-

由Menos é Mais建筑事务所和João Mendes Ribeiro建筑事务所共同设计的大里贝拉Arquipélago当代艺术中心位于亚速尔群岛中的圣米格尔岛。亚速尔群岛通常被看作是欧洲最神秘的群岛。这些小岛屿离葡萄牙海岸1500公里远,几乎位于大西洋的中部,以其美不胜收的风景和如诗如画般的16世纪巴洛克式建筑而闻名于世。

这座当代艺术中心位于一家废弃的烟草和酿酒厂里(廉价国际航运的崛起导致了工厂被废弃),建筑非常美观,融原始的工业建筑(基本都与土地有关)与精密的现代性于一体。粗糙、黝黑、漂亮的火山石与光滑的白色抹灰外墙之间的对比营造了一种张力,相得益彰,而不是相互抵触。为了使其具有现代化的公共事业设备和基础设施,建筑师用黑色的实体混凝土建了一栋新建筑使这个综合体更好地融合在一起。

这座当代艺术中心的功能多种多样:艺术和文化产品创作、仓储设施、多功能厅、表演艺术学校、实验室和艺术家工作室。亚速尔群岛的旅游业虽然起步很慢,甚至在20世纪90年代时还保存着几乎原始未开发的状态,但是现在旅游业却占了其经济的60%。因此对这些岛屿而言,继续吸引游客非常重要,对其目前的经济形势更是如此。虽然亚速尔群岛的失业率比整个葡萄牙的失业率低得多,但是对旅游业的依赖严重削弱了他们对不可控因素的应变能力,主要不可控因素是飞机燃料的成本(并直接影响到机票的价格)。

事实上,由于其特殊的战略地理位置,作为与葡萄牙、北约以及欧盟协定的一部分,亚速尔群岛驻有来自二十多个国家的战斗机。当代艺术中心建设时恰逢亚速尔群岛面临重大的经济困难时期,当时500名美国空军人员和400名葡萄牙工作人员从亚速尔群岛撤离,由此影响到了当地1600个工作岗位,使其GDP也减少了6%。虽然有些建筑看起来像是孤立的机构,构成当地文化产品创作大环境的一部分,然而事实上像当代文化中心这样的项目对亚速尔群岛的生死存亡非常重要。三十年前,渔牧业是亚速尔群岛的未来,如今则变成了美丽而迷人的艺术创造。

由51N4E工作室在科特赖克设计的布达艺术中心充分利用其周边

nowned for their staggeringly beautiful landscapes and picturesque 16th-century Baroque architecture, the small chain of islands sits 1500 kilometres off the coast of Portugal in nearly the middle of the Atlantic.

Housed in an abandoned tobacco and alcohol factory (rendered obsolete by the rise of cheap global shipping), the Contemporary Arts Center is a beautiful complex combining primitive industrial (almost agrarian) architecture with precision modernity. The contrast of beautiful volcanic stone, rough and black, with the smooth white rendered exteriors creates a tension that is complementary, not combative. In order to provide modern utilities and infrastructure, a major new building cast in solid black concrete serves to unite the complex.

The purpose of the Center is diverse: arts and cultural production, storage facilities, a multipurpose hall and performing arts school, as well as laboratories and artist studios. Although the Azores was very slow to embrace tourism – even in the 1990s the island remained virtually pristine – it now accounts for 60% of the economy. It is therefore vital that the islands continue to attract visitors, even more so in its current economic situation: although the Azores have much lower unemployment than Portugal as a whole, their dependence on tourism makes them very vulnerable to factors they can't control, mainly the cost of aircraft fuel (and thus tickets).

In fact, because of its strategic geographic location, the Azores host fighter jets from over 20 countries as part of an agreement with Portugal, NATO and the EU. The timing of the Contemporary Arts Center has been well timed with a major economic problem for the Azores, which is the withdrawal of 500 US Air Force and 400 Portuguese personnel, impacting on a total of 1600 local jobs and cutting the islands' GDP by 6%. Although they may seem like isolated institutions, forming part of a general background of local cultural production, in fact projects like the Contemporary Arts Center are central to the survival of the Azores. Thirty years ago, dairy and fishing comprised their future. Today, it is the production of beautiful and engaging art.

This couldn't be further from the conditions surrounding the Buda Art Center by 51N4E in Kortrijk. Although this is located in Belgium, it is in effect part of Lille's metropolitan area. The importance of this is that Kortrijk is not competing with its compatriots, but with a neighboring French city. Despite the global recession, and being situated in one of the most industrially depressed regions of Northern Europe, unemployment in Kor-

环境,巧夺天工。虽然从地理位置上来说布达艺术中心位于比利时,但是实际上它属于法国里尔大都市区的一部分。布达艺术中心的重要性在于科特赖克并不是与本国其他城市相竞争,而是与其相邻的法国城市一争高下。尽管受到全球经济衰退的影响,又位于经济最为低迷的北欧地区,但是科特赖克的失业率竟惊人地低:只有2%~3%。这座小城市正蒸蒸日上,蓬勃发展,相对富裕,即使与此同时科特赖克正在着手将工业生产项目移出市区也不产生影响。

与亚速尔群岛当代艺术中心的原址闲置了数十年正相反,布达艺术中心的原址一直在使用。这里是科特赖克市最后一家被关闭的纺织厂,现在用很低的成本将它改造成了艺术家工作室和展览空间,成为把以前的工业建筑转化为城市文化活动中心的重要先例。品质不一的原建筑被重新赋予了新的功能而不是重新设计。如今,我们工作和休闲娱乐的方式正在改变,忽然之间,艺术创作也意味着减少屏幕眩光,而不是让空间到处充满着光线。这就意味着通风设计不仅仅要提供户外吸烟平台,也要提供3D打印机的排气管道。

从建筑方面来说,布达艺术中心的主要特色是新建了两座砖塔,补充了原建筑的几何性与物质性。这些元素展现了对工业建筑进行翻修改建时所采用的两个主要建筑策略之一:设计假体。假体是为了增强身体的力量而安装在人体上的一种人造结构(比如说假肢和假牙)。此处的建筑假体即金色的砖砌采光井,在很大程度上或多或少地保留了旧建筑以前的样子,但其设计目的只是改变它的用途。这种另作他用的做法是更加容易让人理解的选择,因为在很大程度上它是可逆的。假体抑制自然蜕变的过程,不会实现蜕变。在这个特别的设计案例中,布达艺术中心可以被理解为从纺织品生产到艺术创作的快速转变,以期使科特赖克市获得比其竞争对手里尔市更多的关注。与假体相对应,对后工业化时期建筑进行重建所采用的另一个主要策略被称为骨架。O-office建筑师事务所就把这种方法用在了满京华美术馆改造上。满京华美术馆位于距离中国深圳市区大约50公里的地方。在这里,鸿华印染厂的旧建筑被看作是一块白板,这个临时混凝土建筑成了一个骨架,其历史痕迹不格外注意的话很难看到。

trijk is almost unbelievably low: just 2%~3%. The small city is thriving, relatively wealthy and growing, even at the same time that it is implementing a program of removing industrial production from its municipality.

The site for the Art Center was never abandoned, unlike in the Azores which remained unoccupied for decades. Rather, it was the last textile factory in Kortrijk to be closed, at which point it underwent an inexpensive transformation into artists' studios and exhibition spaces. It is now an important precedent for the regeneration of the ex-industrial area into the center for cultural activity in the city. The various qualities of the original building have been repurposed rather than redesigned. The ways we work and relax today are changing, and suddenly artistic production means reducing screen glare, not flooding spaces with light. It means ventilation has to be as concerned with 3D-printer exhaust ducts as the provision of outdoor smoking terraces.

Architecturally, the main gesture was the introduction of two new brick towers, complementing the original building's geometry and materiality. These elements demonstrate one of the two main strategies in architecture for treating industrial restoration: as prosthesis. The prosthesis is an artificial structure added to augment the power of a body (as with the prosthetic limb, or dental fixtures). The architectural prosthesis here, the blonde brick light wells, largely leave the original building more or less as it was originally, but simply aiming to alter its functions. This act of appropriation can be seen as the more sensitive option, because it is largely reversible. The prosthesis arrests the natural process of decay; it doesn't achieve a metamorphosis. In this particular case, the Buda Art Center can be interpreted as part of an attempt to very rapidly switch from textile production to artistic production, in the hope of drawing more attention to the city from its rival Lille. Against the prosthesis, the other major strategy for post-industrial restoration might be called the skeleton. This is the way the O-office Architects approached their transformation of the MJH Gallery, situated some 50km from the Chinese city of Shenzhen. Here the original structure of the Honghua Printing and Dyeing Factory was treated as a blank slate. The concrete pavilion became a skeleton, its history visible only in some careful locations.

The symbolic importance of this is very interesting. The factory was built in 1989, at exactly the moment China was globalizing its economy under the reform and opening-up policy initiated

深圳鸿华印染厂改造项目的象征意义非常有趣。这家工厂建于1989年，当时正值中国经济变革之际，也就是在邓小平领导下从20世纪70年代末开始实施的改革开放政策。周边的村庄是深圳外来务工人员最初定居的地方之一。深圳鸿华印染厂代表着中国经济占主导地位的开始，这也就是其改造项目意义非同寻常的原因。曾推动中国进入21世纪的工业生产现在被重新诠释为一个骨架，被现代美术馆造价不菲的室内设计所覆盖。整个项目设计非常优雅，富有想象力。废旧的过滤池也被重新改造为倒影池。

法国的地区当代艺术基金会（FRAC）的设立是源于法国政治家贾克·朗在20世纪80年代早期发起的一项文化活动，是权力下放政策的一部分，旨在提高地方市政当局的组织能力，支持当代艺术的创作和传播。作为这项文化活动的一部分，FRAC北部加来海峡中心最近刚刚在法国北部城市敦刻尔克竣工。该中心由Lacaton & Vassal设计，可以举行临时艺术展览、公共活动，更重要的是，可以进一步增加当地当代艺术品的公共馆藏。这个项目是对该城港口现存的旧船仓库Halle AP2进行翻修改造，并扩建一座具有相同比例和大小的建筑。整个项目的设计旨在尽量扩大工业建筑的规模，促进这座城市的公共和文化生活。虽然和原有建筑Halle AP2的形状相同，但是新建筑所用材料不同，设计也更为复杂。新建筑不仅强调了工厂仓库的标志性外观，而且无论是FRAC内部空间还是周围整体的都市风光，都带给参观者一种全新的空间体验。

由OMA设计的普拉达基金会艺术博物馆位于米兰南部的Largo Isarco，由总占地达19 000m²的几栋建筑组成。像米兰其他工业区如比可卡、波维萨、梅切纳特还有Lambrate-Ventura一样，Largo Isarco是20世纪初意大利工业革命的产物。由几座大型废弃建筑物构成的新基金会项目坐落在城市南部边缘，就在城市轨道高速公路的外围。在原先的Società Italiana Spiriti酿酒厂的基础上，普拉达基金会艺术博物馆由七座原有建筑以及三座新建建筑（矮墙区、电影院和塔楼）组成。这个项目的目的就是通过引入城市的文化和艺术生活来重新提升这一荒废地区的品质。或许最能体现该项目目的的一个地方就是"鬼屋"：一座表

in the late 1970s under Deng Xiaoping. The surrounding village was one of the first settlements to be occupied by immigrant laborers in Shenzhen. It represents the very beginning of Chinese economic dominance, which is why its gutting is potent. The industrial manufacture that propelled China into the 21st century is now reinterpreted as a skeleton, covered over by the expensive gloss of modern gallery interiors. The result is very elegant and imaginative, with the old filtering basins reframed as architectural reflecting pools.

The FRAC (Regional Contemporary Art Funds) originated from a cultural initiative by the French politician Jack Lang in the early 1980s as part of a decentralization policy intended to provide the regional councils with more organizational capacity, to support the creation and dissemination of contemporary art. As part of this initiative, the FRAC Nord-Pas de Calais has been recently completed in Dunkerque, in the north of France. Designed by Lacaton&Vassal, the centre houses temporary art exhibitions, public events, and – more generally – promotes the public collections of contemporary art of the region. The project consists in the rehabilitation of an existing old boat warehouse in the city's port called Halle AP2, which has been coupled by an extension of identical proportions. The compound has been designed to maximise the great scale of the industrial buildings, and act as a catalyst for the public and cultural life of the city. Albeit with the same shape of the existing Halle AP2, the new building has been built with different materials and a more elaborated programme. Not only the new building emphasises the iconic image of the industrial warehouse, but it also offers the visitors new spatial experiences of both the inside of the FRAC and the overall view of the cityscape.

The Prada Foundation Art Museum designed by OMA is located in Largo Isarco, in the south of Milan, and comprises several buildings for an overall surface of 19,000 square metres. As many other industrial areas in Milan like Bicocca, Bovisa, Mecenate, or Lambrate-Ventura, Largo Isarco has been built as consequence of the Italian industrial revolution in the early 1900s. Just outside of the city's orbital motorway, the new Fondazione sits in the south periphery of the city, characterised by several large abandoned buildings. Formerly in the premises of the distillery Società Italiana Spiriti, the Prada Foundation combines seven existing buildings with three new structures: the

面由24k金箔镶嵌的建筑。这座金色建筑从看起来全是一样的灰色建筑物中脱颖而出，充分证明把废弃的工厂转变成新建筑是有潜力的。在这里，艺术和文化与城市建立了新的对话。

由Riegler Riewe建筑事务所设计的、位于卡托维兹市的西里西亚博物馆花费7000万美元巨资建造，证明了当地人民非常为他们的工业遗产而自豪。但是尽管耗资不菲，卡托维兹的人民却花费得起：自16世纪起，卡托维兹就是重要的煤炭和钢铁生产中心。如今，它仍是欧盟中位居第16强的城市，虽仅有270万人口，但年产值却达1140亿美元，是欧洲失业率最低（2%）、波兰月薪最高的城市。毫不奇怪，它的市徽是一把古老的蒸汽锤。

博物馆的设计与这个矿业小镇很相称，大部分位于地下，地面上巨大的玻璃温室为功能区提供光线。这些抽象的形式本身就是高科技工业的产物，展现出了波兰这一最重要的科技、教育、研究中心的巨大力量。因为这个建筑项目是为这座繁荣昌盛的城市所取得的成就加冕，所以其吸引新投资的作用并不会让人一目了然。浮华的作品往往会出现这样的典型问题：因为没有想方设法吸引参观者，这种作品可能一直无人问津，荒凉孤寂。

虽然所有相关机构都普遍关注这一问题，希望这样的事不会发生在卡托维兹市，但是，无论是展示一个城市的文化遗产，还是展示生产方式的根本性转变，甚至是为了吸引参观者，每个项目都有想要引人注意的理由。最终，将是建筑师的设计能力决定一个项目的成败。

Podium, the Cinema and the Torre. The project has the manifest intention to trigger the requalification of the dilapidated area by attracting the cultural and artistic life of the city. Perhaps the part of the project that illustrates this aim at the most is the Haunted House: a tower clad in 24-carat gold leaf. Standing out from the rest of the compound which appears as a homogenous grey background, this golden tower epitomises the potential which lies behind the transformation of abandoned industrial structures into a new richness where arts and culture can establish a new dialogue with the city.

At a staggering cost of $70 million, Silesian Museum in Katowice by Riegler Riewe Architekten is a testament to the pride the city's people take in their industrial heritage. But then, they can afford it: since the 16th century Katowice has been an important center for coal mining and steel production. Today it is the 16th most powerful city in the EU, generating $114 billion in annual output for a city with a population of just 2.7 million. It has the lowest unemployment in Europe (2%) and has the highest monthly salary in Poland. Unsurprisingly, its coat of arms is an archaic steam hammer.

The Museum, appropriately for a mining town, is mostly underground, with the vast glass greenhouses on the surface drawing light deep into the plan. These abstract forms, which could easily be high-tech industrial artifacts themselves, demonstrate the immense power of Poland's most important center for science, education and research. Because the project serves to crown the achievements of what is already a prosperous city, its role in attracting new investment is not clear. There is a typical problem that accompanies vanity works – they are not desperate to draw in visitors and can remain unfrequented or desolate.

Hopefully this will not become the case in Katowice, although this is a general concern for all these institutions. Whether to show off a city's heritage, or demonstrate a fundamental shift in modes of production, or even with the urgent desire to attract tourists, each project has its own reasons for wanting to be noticed. In the end it is the skill of the architects that is instrumental in this success. *Jack Self*

布达芝术中心
51N4E

作为新型产业的艺术 Art as the New Industry

布达岛位于科特赖克市人口稠密的市中心位置,这里充分体现了这座城市丰富而独特的历史风貌。近十个世纪以来,这个小岛不断发展变化,从一个为穷人和病人提供庇护的天主教庇护所,发展成为依赖水资源的繁荣的制造业基地(啤酒厂、印染厂、漂白厂)。奇怪的是,上述两种截然不同的历史命运今天仍然以独特的方式留下了痕迹:过去的庇护所已经让位给了医院和养老院,过去的工业厂房也被改造成了文化设施。

因此,布达岛上弥漫着不受拘束的甚至是随便的氛围,任何事都充满实验性。科特赖克市致力于将自身逐步打造成一座"设计之城",不放过任何在国际设计舞台上留下浓墨重彩的机会。

这一努力的最新体现就是布达工厂。布达工厂是一座建于1924年的纺织印染厂,本身没有任何特别引人注目的空间特点或历史特色,唯一的设计挑战就是以极低的成本为广泛意义上的文化表现提供庞大而宽敞的空间,而内部如何利用则被认为是次要的。

布达工厂的改造强调利用适当的基础,保留了过去的一个生产制作车间。为了使目标空间最大化,空间尽可能宽敞,大部分工厂的空间被重新分割、改造,使空间既精致又绝对宽敞。

两个中空的五边形空间,分别位于室外和室内,形成光线、规模和随意性的内化序列,里外呼应。外面的这个五边形体量形成了一个前厅,模棱两可的存在感使其与两边一列列紧凑的建筑格格不入。里面的五边形体量将一系列工作室、展示空间和工作间连为一体,直通屋顶平台,与周边环境和科特赖克市的城市风光相连。

总之,布达工厂是用来观看世界的一个工具,而不是用来观看的物体,集合了格格不入的已发现空间和新增空间,待众多形形色色的观众去探究他们想要探究的一切。

Buda Art Center

The Buda island encompasses the dense city heart of Kortrijk, mirroring the city's rich but idiosyncratic history. Since the last ten centuries this insular territory has evolved from a Catholic shelter environment for the poor and the sick into a prosperous industry for water-dependent manufacturing (breweries, dye houses, bleach plants). Oddly enough these two radically different historical destinies are still extremely present today, though the shelters have been reformatted into hospitals and infrastructure for the elderly, while the industrial witnesses have found themselves turned into urban envelopes for new cultural programs.

Hence an atmosphere of unbound – even haphazard – experiment lingers over the Buda island. In its endeavor to gradually excel as the City of Design, Kortrijk spares no opportunity to score on the international design ladder.

The latest evidence of this striving is the Buda factory, a textile

项目名称:Buda Art Center / 地点:Kortrijk, Belgium
建筑师:51N4E
客户:City of Kortrijk
功能:artist studios, exhibition facilities
用地面积:3,900m² / 总建筑面积:4,240m²
造价:EUR 2,545,000 (excl. VAT)
竞赛时间:2005 / 竣工时间:2012
摄影师:
©Filip Dujardin (courtesy of the architect) - p.10~11, p.14, p.15, p.16, p.17[top]
©Paul Steinbrueck (courtesy of the architect) - p.12, p.17[bottom], p.18, p.19

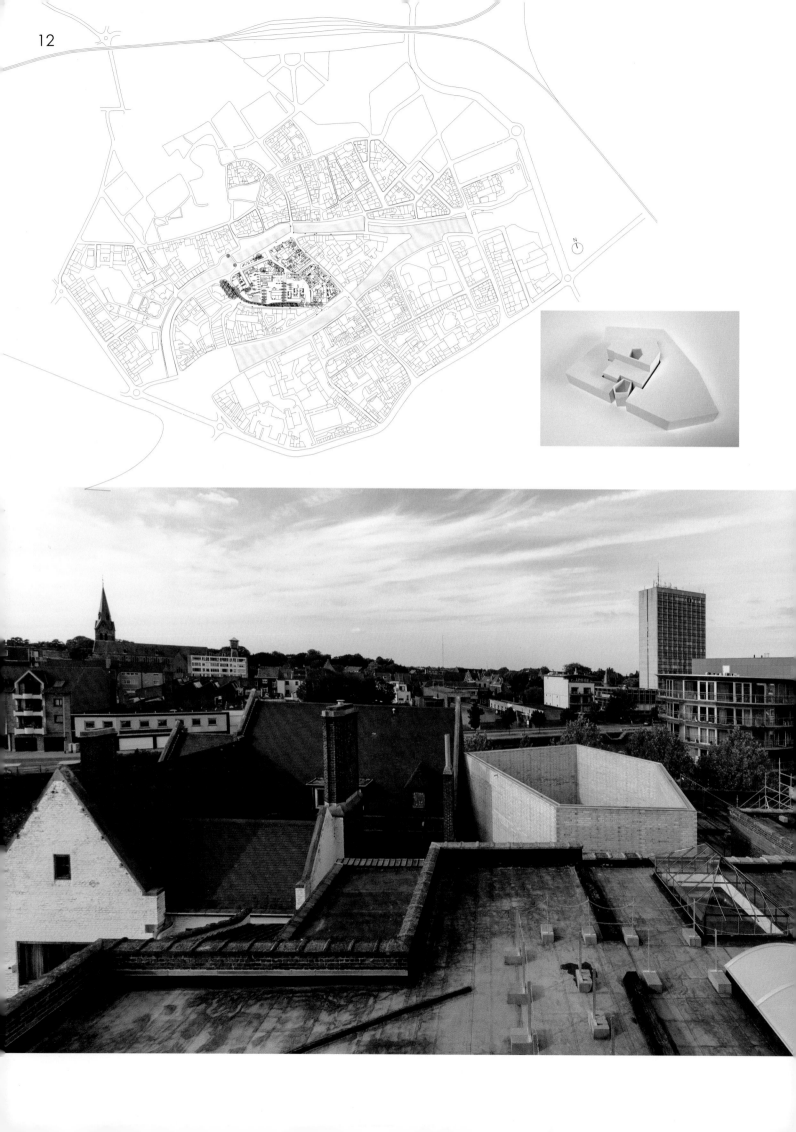

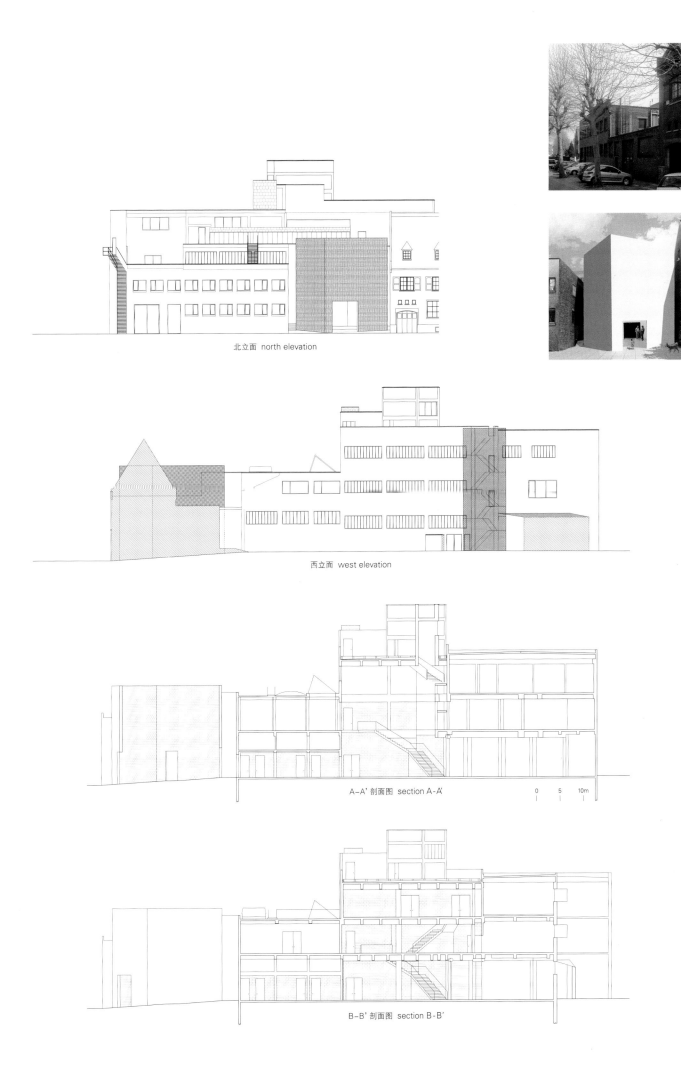

北立面 north elevation

西立面 west elevation

A-A' 剖面图 section A-A'

B-B' 剖面图 section B-B'

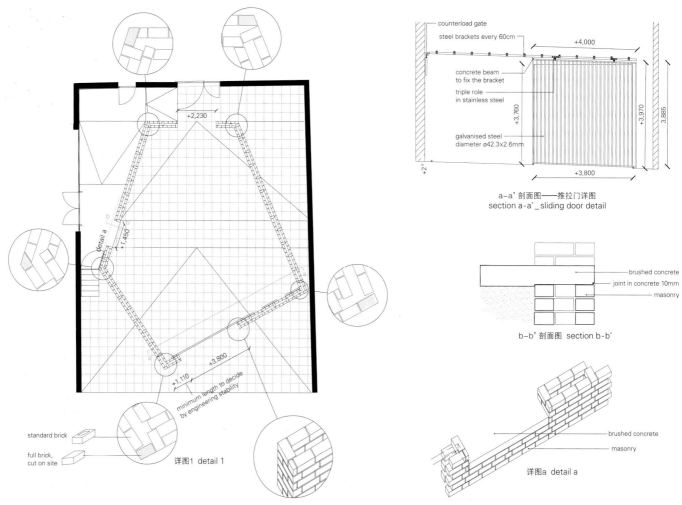

dye plant originating from 1924, with no particular worthwhile spatial or even historical features. The sole challenge of the Buda factory is to offer an extremely large and generous space for cultural expression in its broadest meaning and built at an extremely low cost. The programmatic infill is consciously considered as secondary.

The refitting of the Buda emphasizes its proper foundation, referring to its past by simply remaining a workshop for production. Most of the factory's spaces are re-trimmed and reconfigured as to cater for a maximum of objective and excessive space, delicate and brutal at once.

Two hollow pentagonal spaces, one outside and one inside, introduce an interiorized sequence of light, scale and indeterminacy. The outside volume performs as an antechamber, rupturing with its ambiguous presence the tight row of buildings that frame it. The inside pentagon juxtaposes a series of studios, exhibition spaces and workshops, cumulating into a rooftop terrace as the sole encounter with the context and the cityscape of Kortrijk.

In brief, the Buda factory is a tool to look with, not an object to look at. It is a collection of found and added spaces – of an unadapted nature – allowing for innumerably diverse audiences to probe for whatever they are looking for.

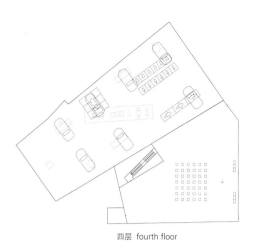

四层 fourth floor

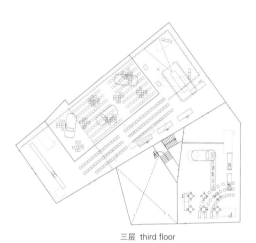

三层 third floor

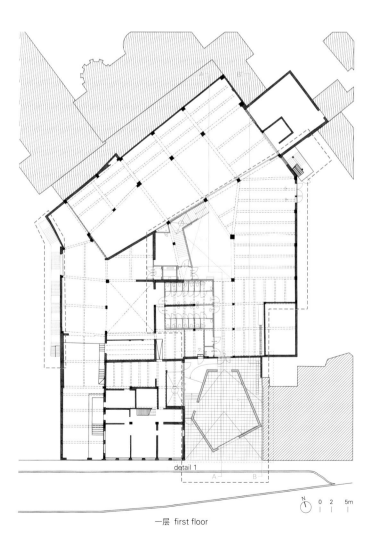

一层 first floor

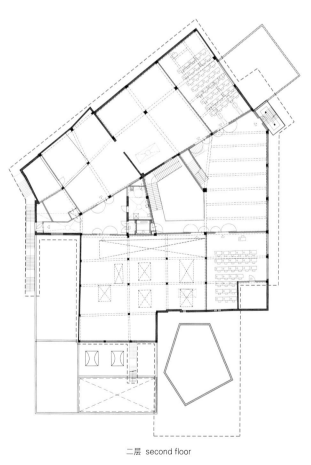

二层 second floor

Arquipélago当代艺术中心致力于将其各个部分的不同规模与时代特色统一起来,这是一个跨学科项目,其使命是传播、创作和制造新兴文化:为人们交流、传播知识、举办活动营造一个空间和场所。

Arquipélago当代艺术中心的设计保留了整个工业特征,突出原有建筑(之前的酒厂/烟草厂)和新建筑(艺术和文化中心、仓储设施、多功能大厅和表演艺术区、实验室、艺术家工作室)之间的对话。

Arquipélago当代艺术中心通过原有建筑和两座新建筑之间的平和过渡变化获得认同。设施之间相互牵制的设计战略提高了原有工厂不同区域的空间效率和层次功能。新建筑吸收了所需的功能,因其特殊条件,与原有建筑的空间性并不兼容。Arquipélago当代艺术中心项目没有夸大新旧建筑之间的差异。相反,通过对所有建筑的形式和材料进行图像操作处理,致力于将其组成部分的不同规模与时代特色统一起来:原有建筑的特点是火山岩砌石,新建筑的特点是其抽象的建筑形式,没有参考或暗指任何建筑语言,由当地玄武岩混凝土建成,使整个项目形成变化的表面纹理和褶皱,以天井的空洞感来补充整体建筑的实体感。此设计致力于突出原有建筑物的品质,展示了建筑类型的变化——新建筑放置在原有建筑旁——既凸显了人们对特定时期建筑的记忆和新增加的建筑,又没有破坏或颠覆整体的空间和构造上的结构,周围环境均有助于实现建筑的自主性。

这个新项目彻底改造了原建筑,使之成为位于大西洋中部的周边地区中一个富有意义的空间。

Arquipélago当代艺术中心使场地的社会和文化环境具有了新的意义。中心广场/天井内建造了一个新的公共空间,这一空间与整个艺术氛围毫无违和感,使私人领域与公共领域、休闲与工作、艺术与生活之间的边界变得模糊。

建筑的可持续性能方面是通过其物质性(结构、基础设施)和利用经久不衰的传统手工艺技术来实现的。所采用的可持续措施包括努力为用户提供舒适性的被动系统:混凝土墙的密度提供了惯性和能源效率;雨水可以重复使用。

Arquipélago当代艺术中心
Menos é Mais Arquitectos + João Mendes Ribeiro Arquitecto

Arquipélago Contemporary Arts Center

The Arquipélago Contemporary Arts Center seeks to unite the different scales and times of its parts. It is a transdisciplinary project whose mission is to disseminate, create and produce emerging culture: a space of exchange and interface for people, knowledge and events.

The design of the Arquipélago maintains the industrial character of the whole and highlights the dialogue between an existing building (former factory of alcohol and tobacco) and the new construction (arts and culture center, storage facilities, multipurpose hall and performing arts areas, laboratories, artist studios).

The Arquipélago acquires its identity by the quiet variation between the preexistence and the two new buildings. The containment strategy of facilities implementation enhances the spatial efficiency and hierarchical functionality of the different areas of the existing factory complex. The new buildings absorb the required functionality, with special conditions, not compatible with the spatiality of preexisting buildings.

The project of the Arquipélago does not exaggerate the differences between the old and the new buildings. On the contrary, it seeks to unite the different scales and times of its parts throughout a pictorial manipulation of the form and materiality of the buildings – the existing constructions are marked by the volcanic stone masonry and the new buildings are characterized by an abstract form, without reference or allusion to

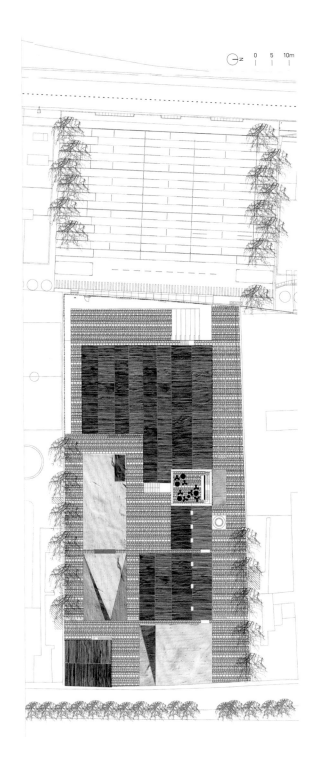

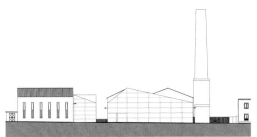

南立面 south elevation

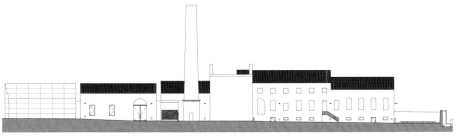

东立面 east elevation

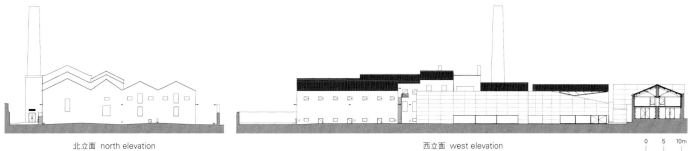

北立面 north elevation　　　　西立面 west elevation

三层 second floor

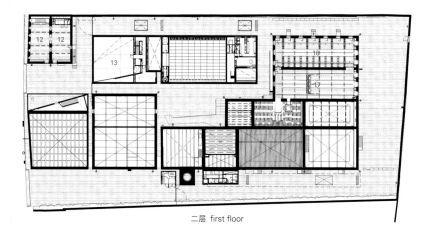

二层 first floor

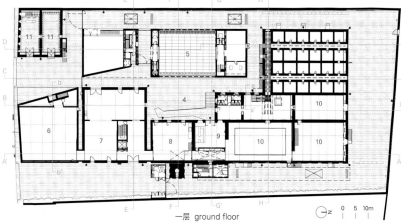

一层 ground floor

1 工艺与艺术协作区
2 技术区
3 多媒体区
4 门厅
5 礼堂和多功能区
6 储藏室
7 工作坊和木工室
8 展览装配与拆卸区
9 入口大厅与接待处
10 展厅
11 博物馆商店与书店
12 工厂模式的博物馆中心
13 工艺与艺术支持区和实验室
14 隔声工作室
15 酒吧
16 行政服务区、办公室与会议室
17 教育服务室
18 文档与图书馆中心

1. technical and artistic assistance area
2. technical area
3. multimedia area
4. foyer
5. auditorium and multipurpose area
6. storage
7. workshop and carpentry
8. exhibition assembly and disassembly area
9. entrance hall and reception
10. exhibition
11. museum shop and bookstore
12. factory museum center
13. technical and artistic support area and laboratories
14. soundproofed studio
15. bar
16. administrative services, office and meeting room
17. education services
18. documentation and library center

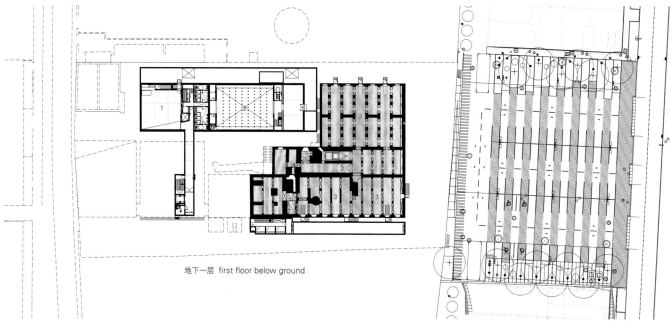

地下一层 first floor below ground

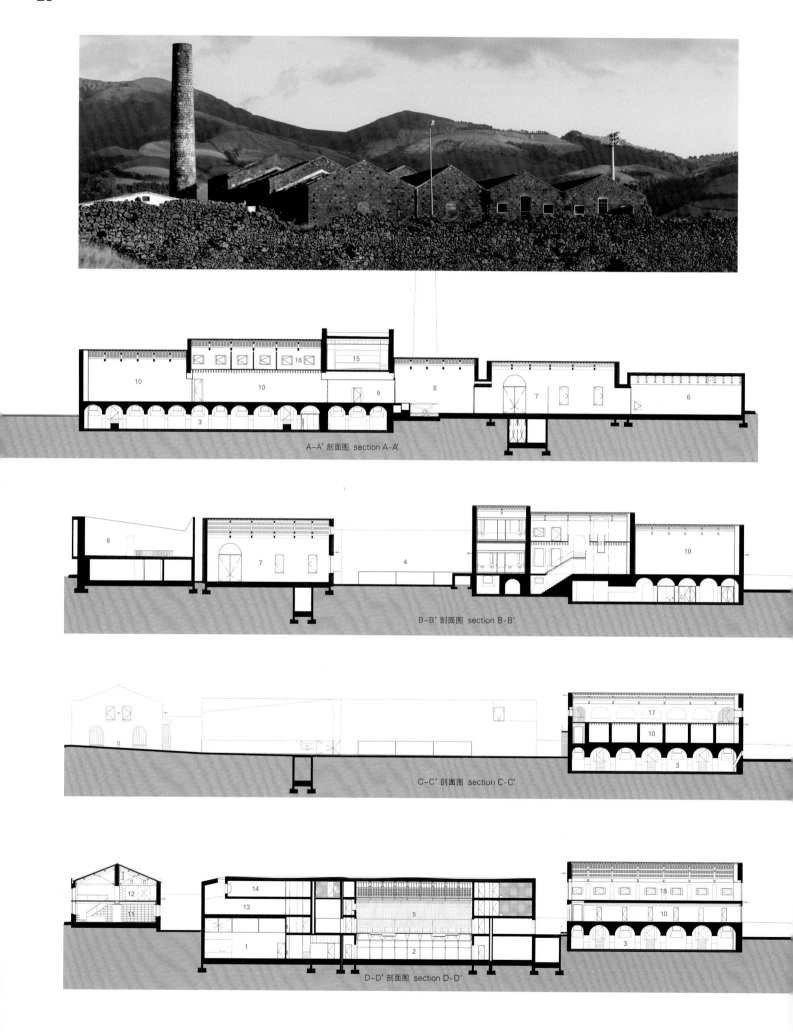

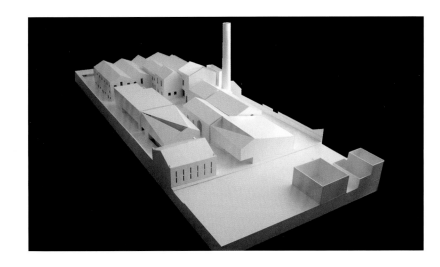

1 工艺与艺术协作区
2 技术区
3 多媒体区
4 门厅
5 礼堂和多功能区
6 储藏室
7 工作坊和木工室
8 展览装配与拆卸区
9 入口大厅与接待处
10 展厅
11 博物馆商店与书店
12 工厂模式的博物馆中心
13 工艺与艺术支持区和实验室
14 隔声工作室
15 酒吧
16 行政服务区、办公室与会议室
17 教育服务室
18 文档与图书馆中心

1. technical and artistic assistance area
2. technical area
3. multimedia area
4. foyer
5. auditorium and multipurpose area
6. storage
7. workshop and carpentry
8. exhibition assembly and disassembly area
9. entrance hall and reception
10. exhibition
11. museum shop and bookstore
12. factory museum center
13. technical and artistic support area and laboratories
14. soundproofed studio
15. bar
16. administrative services, office and meeting room
17. education services
18. documentation and library center

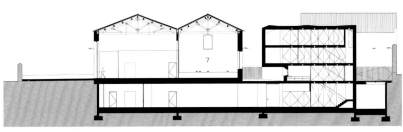

E-E' 剖面图 section E-E'

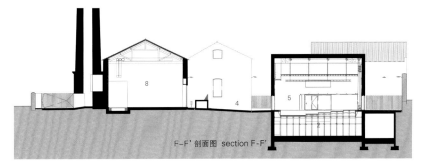

F-F' 剖面图 section F-F'

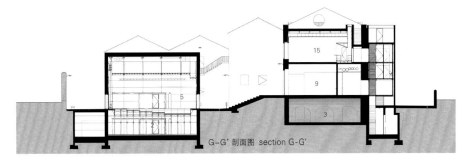

G-G' 剖面图 section G-G'

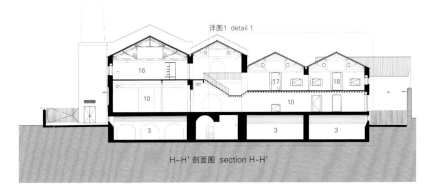

详图1 detail 1

H-H' 剖面图 section H-H'

any language, built in concrete with local basalt continuously working with the variation of surfaces' textures and rugosity, complementing the mass of the buildings with the emptiness of the patios. The design is committed to the quality of what exists, showing the typological variations – new buildings are placed next to the existing ones – underlining the architectonic memory of a given period and the new addition, without damaging or subverting the spatial and constructive structures of the whole. Context and contiguity contribute to the autonomy of the object.

The new program reinvents the existent building, making it a meaningful space in a peripheral region in the middle of the Atlantic Ocean.

The Arquipélago adds meaning to the social and cultural context where it is built. A new Public Space is materialized in a central square/patio where art feels comfortable and blurs the frontiers between private and public spheres, leisure and work, art and life.

The aspects of the sustainable performance of the buildings were addressed through its materiality (structures, infrastructures) and the absorption of the existing handcrafted knowledge enriched by its timeless way of building. The sustainable measures adopted are passive systems that seek to provide comfort for the users: the density of the concrete walls offer inertia and energy efficiency; the rain water is reused.

项目名称：Arquipélago – Contemporary Arts Center
地点：Ribeira Grande, São Miguel, Azores, Portugal
建筑师：Menos é Mais Arquitectos Associados, Lda., João Mendes Ribeiro Arquitecto Lda.
作者：Francisco Vieira de Campos, Cristina Guedes, Joao Mendes Ribeiro
项目经理：Adalgisa Lopes, Jorge Teixeira Dias, Inês Mesquita, Filipe Catarino
项目团队：Cristina Maximino, Joao Pontes, Luis, Campos, Ana Leite Fernandes, Mariana Sendas, Pedro Costa, Ines Ferreira, Joao Fernandes_Menos e Mais Arquitectos Associados, Lda. / Catarina Fortuna, Ana Cerqueira, Ana Rita, Martins, Antonio Ferreira da Silva, Claudia Santos, Joana Figueiredo, Joao Branco_ João Mendes Ribeiro Arquitecto, Lda.
图像后期制作、完成：Diogo Laje, Óscar Ribas, Ricardo Cardoso_Estúdio Goma
结构工程师：Hipólito Sousa, Jerónimo Botelho, Pedro Pinto_SOPSEC,SA
液压装置：Diogo Leite, Filipe Freitas, Jorge Rocha_SOPSEC,SA
电气设备：Raul Serafim, Hélder Ferreira_Raul Serafim & Associados, Lda
安全措施/防火咨询：Maria da Luz Santiago_Raul Serafim & Associados, Lda
机械设备：Raul Bessa, Ricardo Carreto_GET, Lda.
燃气设备：José Pinto_SOPSEC,SA
功能设计咨询：Elisa Babo_Quaternaire, Miguel Von Haff Pérez, Marta Almeida
保护/存储设计咨询：Gabriella Casella_Cariátides
声效咨询：Rui Calejo, Eduarda Silva, Filomena Macedo_SOPSEC, SA
热能咨询：André Apolinário_SOPSEC, SA
舞台机械设计咨询：João Aidos
景观设计：Ana Barroco, Rui Figueiredo_Quaternaire
承包商：Consórcio Somague, Marques S.A. e Tecnovia.
监理：Pedro Câmara_Eng. Tavares Vieira, Lda.
模型：Menos é Mais Arquitectos Associados, Lda.
客户：Regional Directorate of Culture (DRaC) of the Regional Government of the Azores
用地面积：11,167m² / 建筑面积：3,647m² / 总建筑面积：10,252m²
竞赛设计：2007 / 设计时间：2007—2010 / 施工时间：2011—2014
摄影师：©José Campos (courtesy of the architect)

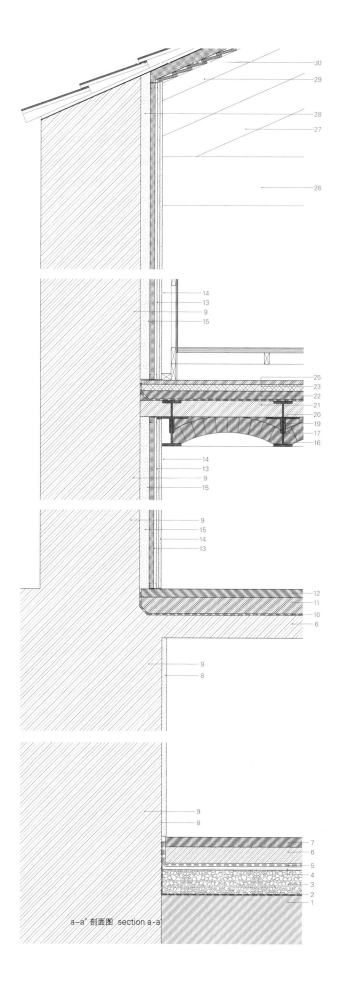

a-a'剖面图 section a-a'

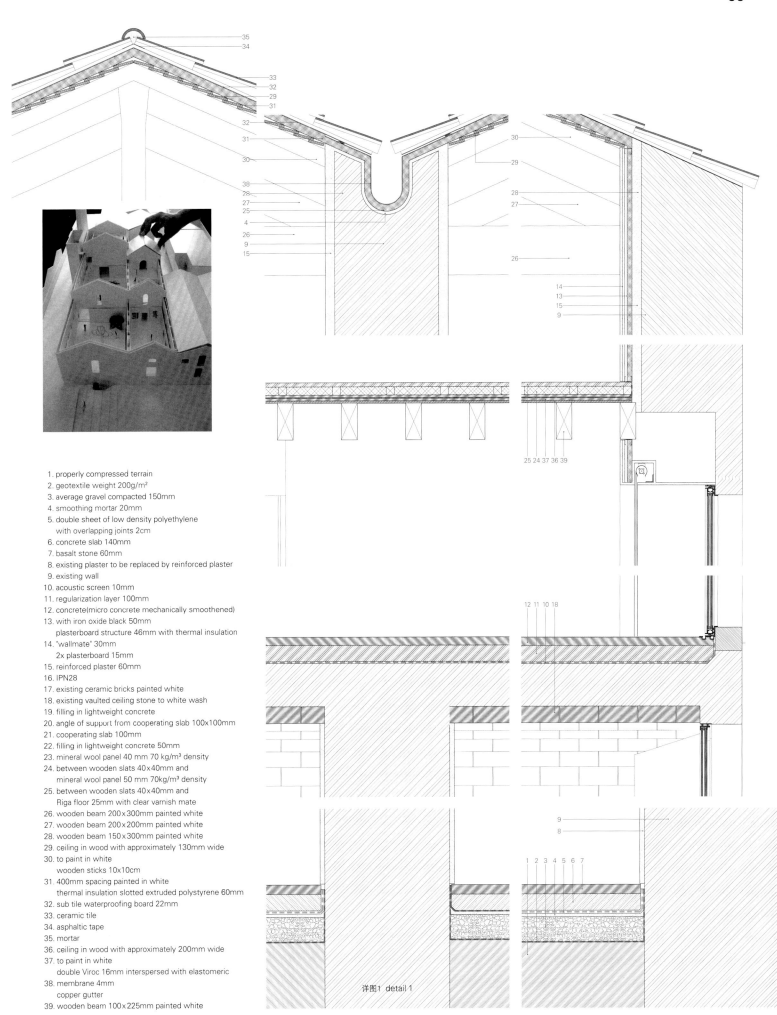

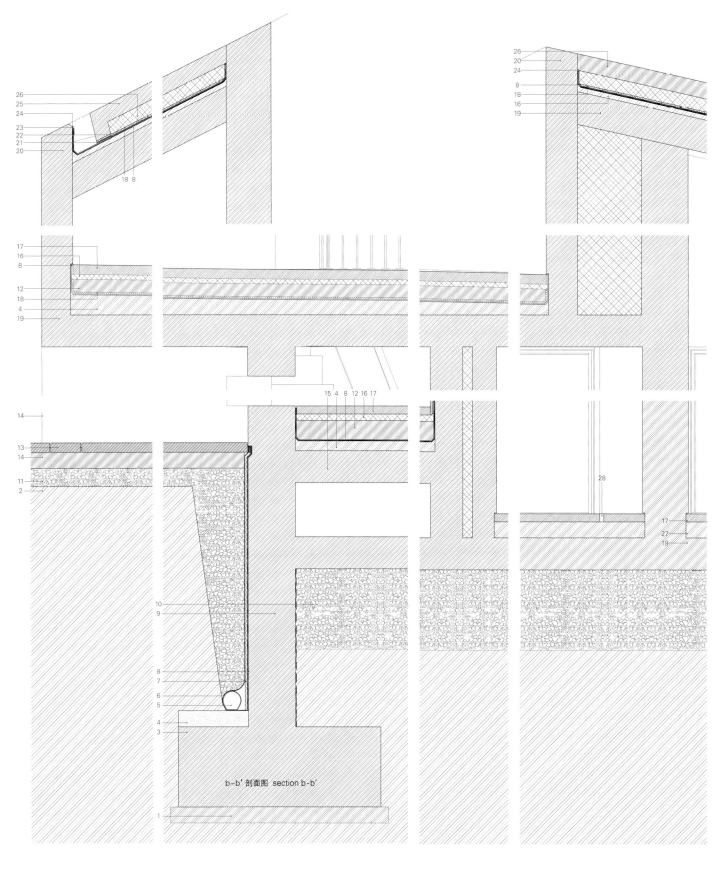

1. lean concrete
2. properly compressed terrain
3. strip footing
4. screed laid to falls 1.5%
5. drain 16mm
6. geotextile grammage 200g/m²
7. draining blade
8. waterproof polyester membranes (3+4mm)
9. concrete 300mm
10. gravel 500mm
11. average gravel compacted
12. concrete slab 100mm
13. basalt stone 60mm
14. concrete disabled polished wall and ceiling surface
15. concrete
16. mortar
17. concrete (micro concrete mechanically smoothened) with iron oxide black 50 mm
18. protective layer
19. concrete slab 200mm
20. deactivated concrete 300mm
21. waterproof polyester membrane 4mm
22. waterproof polyester membrane 3mm

23. ledge profiles
24. elastic mastic
25. concrete slab stapled 100mm
26. thermal insulation extruded polystyrene 80mm
27. filling in lightweight concrete 30mm
28. expansion joint 25 x 50mm

iD Town 满京华美术馆
O-office Architects

满京华美术馆的前身是鸿华印染厂的整装车间厂房,其设计构想源于对这栋空置厂房空间的重新叙事组织。新建筑被单独植入旧厂区的内部,即将原厂房框架看作一个巨大、开放的混凝土展馆,以便把一个新的、多重的空间/时间逻辑融入到工业遗迹中去。

主展厅作为美术馆内部核心空间,为悬置在旧厂房中心位置的一个黑色钢盒体,上方双屋顶的天窗设计非常特别,为展厅提供了柔和的顶部漫射光。这个封闭的盒体展厅通过两个半玻璃大厅与地面相连。几个独立的功能模块一方面支撑着被抬高的主要大厅,另一方面又定义了首层空间和不同的功能模块(包括入口展厅、礼堂、艺术品商店和可以灵活间隔的多功能空间)。首层空间的各个方向都与原来厂区的地面相连,并与周围的自然风光融为一体。

美术馆的主入口设于厂房南侧山墙一侧的原入口处,利用一个T形半开放的混凝土嵌入结构与被改造为景观水池的原印染厂的过滤池相连。建筑师从地形中建出一系列景观墙、人行道和车辆坡道,这样的设计旨在将美术馆的主入口和与郊区道路连通的较低的广场相连。人们沿着这条广场景观路径可以步入美术馆,使参观者仿若穿行于自然与人造物、工业与后工业景观之间,体验和铭记这个地方的整个历史脉络。

MJH Gallery of iD Town

Formerly the packing workshop of Honghua Dying Factory, MJH Gallery was conceptualized from re-structuring the spatial narratives of this vacant building. The new architectonics was independently implanted onto the original inner ground of the old workshop, which was regarded as a large open concrete pavilion, in order to build the new multi-layered spatial/time logic into the industrial relic.

The inner core space of the gallery, the main exhibition hall, is lifted up in a black steel box in the center of the old building, under its special double-roof skylight, which gives gentle diffused top-light to the exhibition space. The enclosed exhibition box connects to the ground floor via two semi-glass halls. Several free-standing functional blocks, on the one hand, support the elevated main hall, and on the other, define the

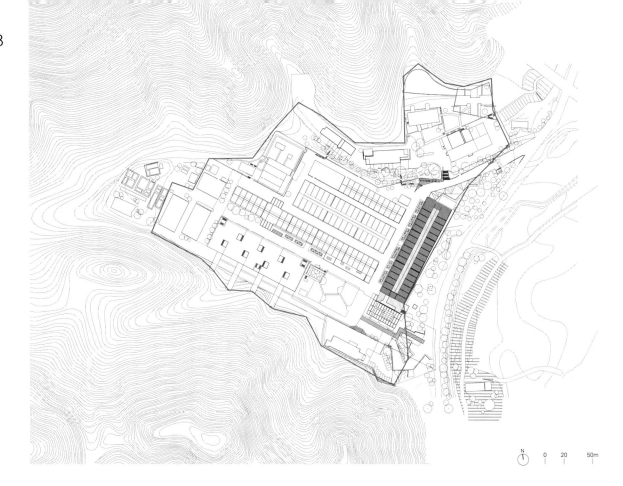

项目名称：MJH Gallery of iD Town
地点：Kuipeng Road 106, Dapeng New District, Shenzhen, China
建筑师：O-office Architects
总建筑师：Jianxiang He, Ying Jiang
设计团队：Jingyu Dong, Thomas Odorico, Xiaolin Chen, Guixiong Feng
原建筑占地面积：2,944m²
新建筑占地面积：2,039m²
总建筑面积：2,800m²
设计时间：2013
施工时间：2014
竣工时间：2014.11
摄影师：©Chaos.Z (courtesy of the architect)

ground space and functions that contain an entry show room, an auditorium, an art shop and related divideable multi-functional spaces. The ground floor is open in all directions to the original ground of the factory, as well as to the surrounding natural landscape.

The main entrance of the gallery, with a half-open T-shape concrete box plugged-in, uses the original entrance on the south gable of the workshop, confronting a landscape pool that converted from the abandoned filtering basin of the factory. A sequence of landscaping walls, pedestrian and vehicle ramps cultivated from the topography was designed to link the gallery's main entrance to the lower plaza that connects the suburban road. This landscape path from the plaza to the gallery detours between nature and artefacts, industrial and post-industrial landscape, through which, one could experience and memorize the whole time line of the site's history.

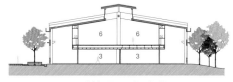

1 入口   1. entrance
2 接待大厅 2. reception hall
3 放映厅   3. video hall
4 礼堂   4. auditorium
5 艺术商店 5. art store
6 展厅   6. exhibition hall

A-A' 剖面图 section A-A'

B-B' 剖面图 section B-B'

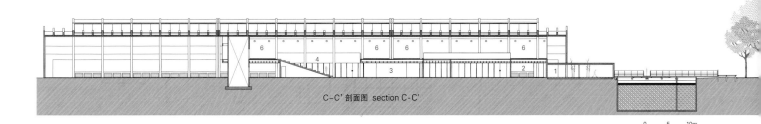

C-C' 剖面图 section C-C'

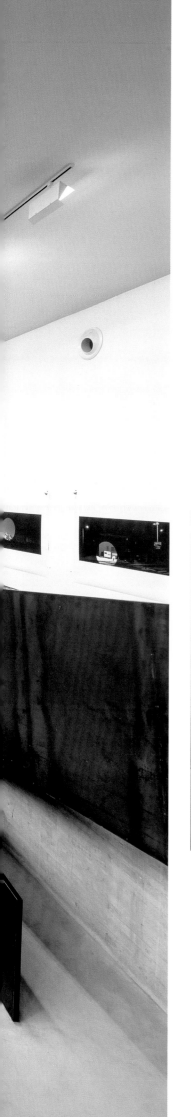

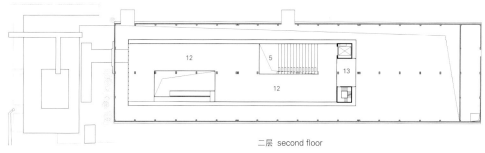

二层 second floor

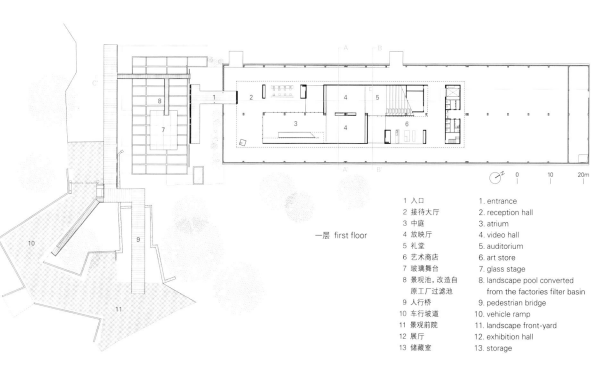

一层 first floor

1 入口	1. entrance
2 接待大厅	2. reception hall
3 中庭	3. atrium
4 放映厅	4. video hall
5 礼堂	5. auditorium
6 艺术商店	6. art store
7 玻璃舞台	7. glass stage
8 景观池,改造自原工厂过滤池	8. landscape pool converted from the factories filter basin
9 人行桥	9. pedestrian bridge
10 车行坡道	10. vehicle ramp
11 景观前院	11. landscape front-yard
12 展厅	12. exhibition hall
13 储藏室	13. storage

普拉达基金会艺术博物馆
OMA

普拉达基金会的新家位于米兰南部的Largo Isarco工业园区里于1910年建的杜松子酒厂的旧址上,是新建筑和改造建筑的融合体,其中包括仓库、实验室和酿造筒仓等旧建筑以及围绕大庭院的新建筑。

该综合体旨在扩大可以进行艺术展览的空间类型。整个项目包括七座现存建筑和三座新建建筑。三座新建建筑是:矮墙区——临时展览的场所;电影院——多媒体礼堂;塔楼——共有9层,用于展出基金会的永久性收藏和举行活动。塔楼还正在建设中,晚些时候会对公众开放。

位于Largo Isarco的这个综合体有两个独立的结构,一个是扁平的方形结构,另外一个是垂直结构。近距离观察,方形建筑毫无吸引力可言,所以被拆除,使露天庭院成为重要的建筑元素。原来位于西侧的建筑——Deposito,比较适合博物馆管理者对于独创性的要求:在其地下室,基金会的收藏品采用混合方式摆放,既严格封存又有部分展示空间,形成了一个个"房间",在这里,像艺术家的车这样的展品可以不加任何包装或半开放面向公众。

大礼堂东面的独立结构被称作Cisterna,共有三个房间,每个房间都有"讲道台",讲道台与外面的阳台相通。这种结构充分展示了如今行业对半宗教环境的需求。

电影院是整个建筑群中的一个独立单元。人们打开大的双开门,就能直接进到院子里。在电影院内,倾斜的座位区可以转换调整为平坦的地面。这样的话,这一空间也可以用来举行户外活动,或作为一个额外的有顶画廊。

四座北朝庭院,南朝一个废弃花园的"建筑"内设基金会的办公室和展览永久藏品的画廊。不远处就是"鬼屋",这栋原有建筑物的整个外立面被涂上了一层金箔。在里面,狭小的室内空间为一些特定的作品创造出"温暖如家"的背景。

旁边是矮墙区,是整座综合建筑的中心,位于两条垂直轴线交叉的地方。这座扩建建筑由两种材质完全不同的建筑体量组合而成:地面一层矮墙区的外立面全是玻璃,里面采用无柱空间设计;上面一层画廊的外立面用泡沫铝覆盖,带有泡泡形穿孔图案。两个画廊都为举办临时展览和重大活动提供了具有多重用途的宽敞空间。

综合体中另外一个主要增建建筑是位于西北角的9层塔楼。这里有餐厅、接待处和一些客用设施,还有一些特定现场装置。随着高度的逐渐增加,其他楼层也会放置一些特定现场装置。另外,参观者在此可以登高眺望到一个不一样的普拉达基金会艺术博物馆和一座不一样的城市。

Prada Foundation Art Museum

Located in a former gin distillery dating from 1910 in the Largo Isarco industrial complex on the southern edge of Milan, the new home of Fondation Prada is a coexistence of new and re-generated buildings including warehouses, laboratories and brewing silos, as well as new buildings surrounding a large courtyard.

The complex aims to expand the repertoire of spatial typologies in which art can be exhibited. The project consists of seven existing buildings, and three new structures: Podium, a space for temporary exhibitions; Cinema, a multimedia auditorium; and Torre(tower), a nine-story permanent exhibition space for displaying the foundation's collection and activities. Torre, currently undergoing construction work, will be open to the public at a later date.

Within the perimeter of the Largo Isarco complex existed two freestanding structures: one flat and square and the second more vertical. On close inspection, the square building did not offer attractive possibilities and was demolished, enabling the courtyard to become a significant element for open-air use. The Deposito, an existing building on the west edge of the complex, is adapted for curatorial ingenuity: in its basement,

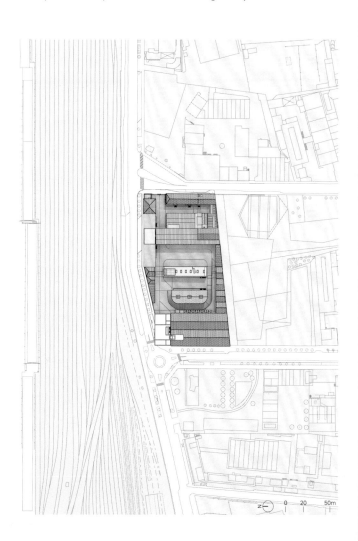

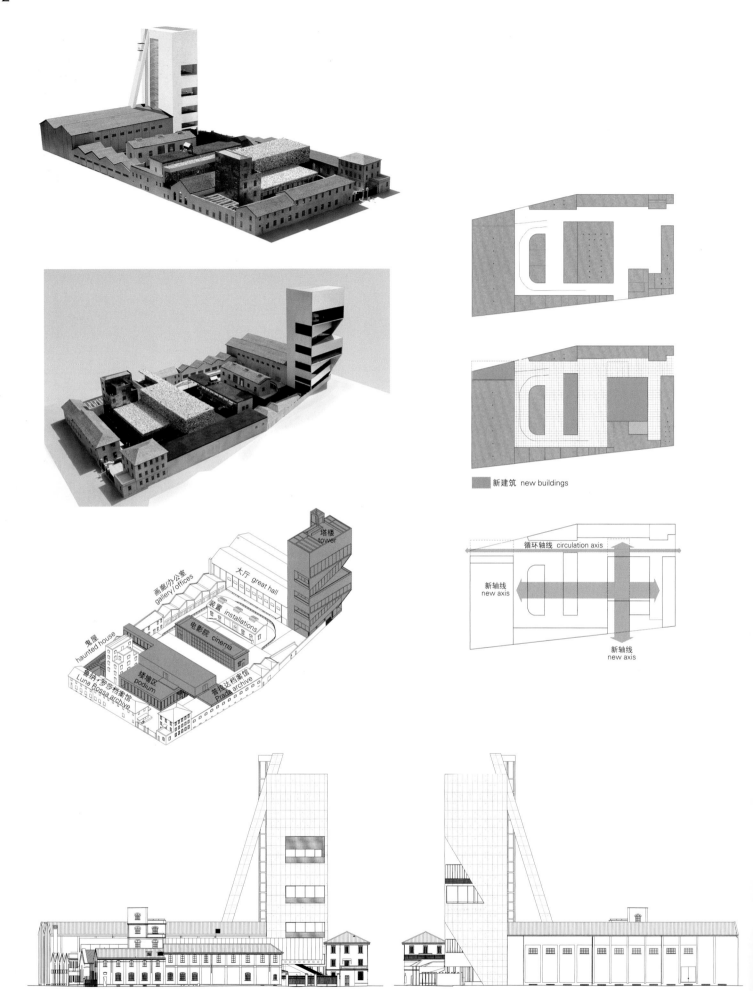

东立面 east elevation   西立面 west elevation

南立面 south elevation

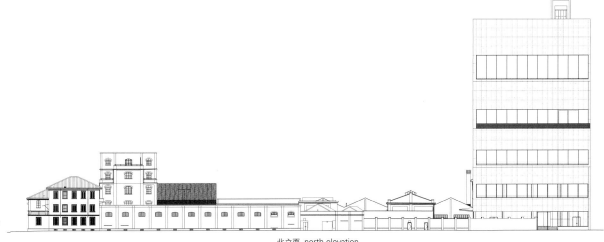

北立面 north elevation

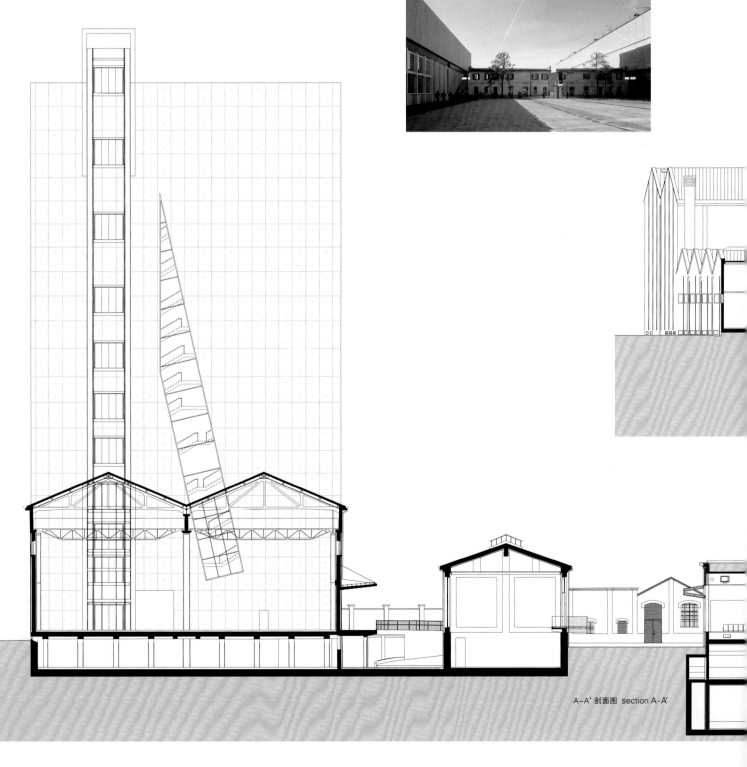

A-A' 剖面图 section A-A'

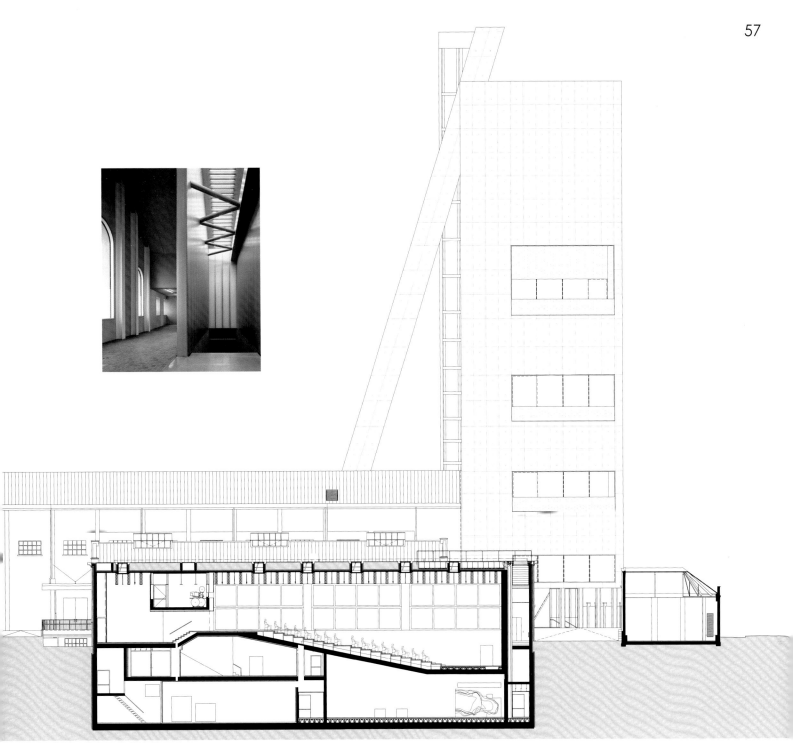

B-B' 剖面图 section B-B'

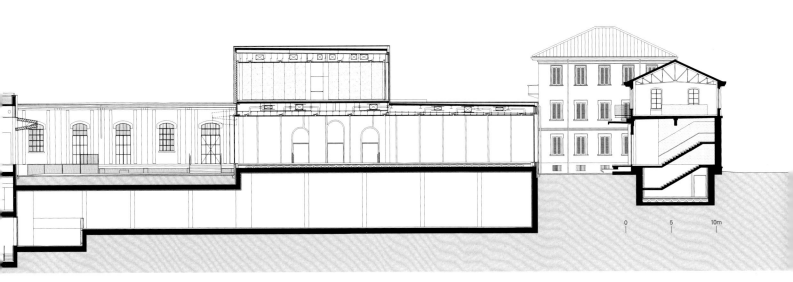

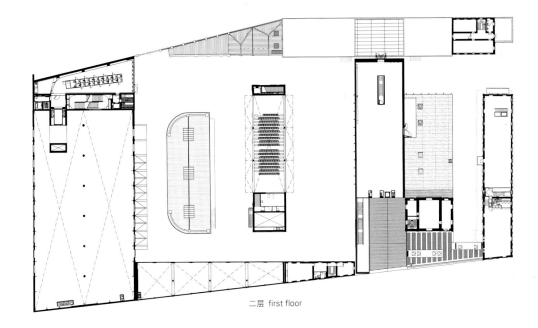

二层 first floor

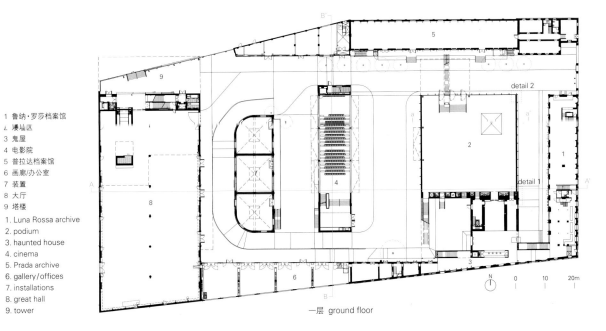

1 鲁纳·罗莎档案馆
2 嫚坛区
3 鬼屋
4 电影院
5 普拉达档案馆
6 画廊/办公室
7 装置
8 大厅
9 塔楼

1. Luna Rossa archive
2. podium
3. haunted house
4. cinema
5. Prada archive
6. gallery/offices
7. installations
8. great hall
9. tower

一层 ground floor

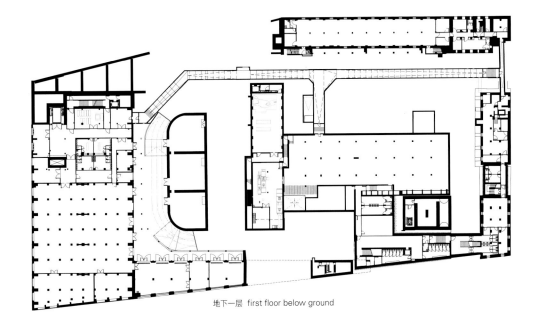

地下一层 first floor below ground

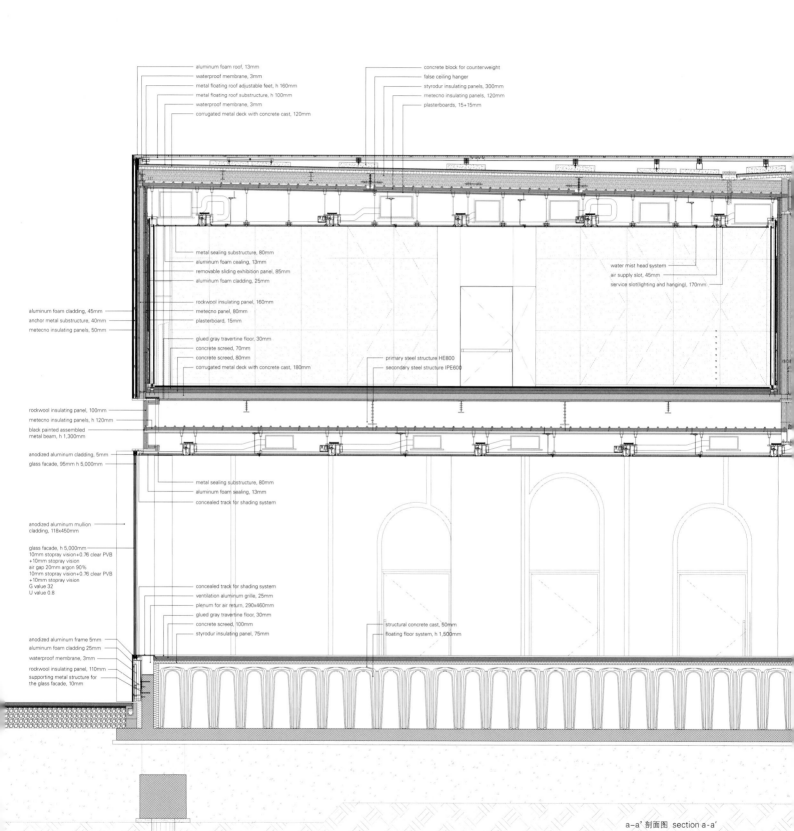

a-a' 剖面图 section a-a'

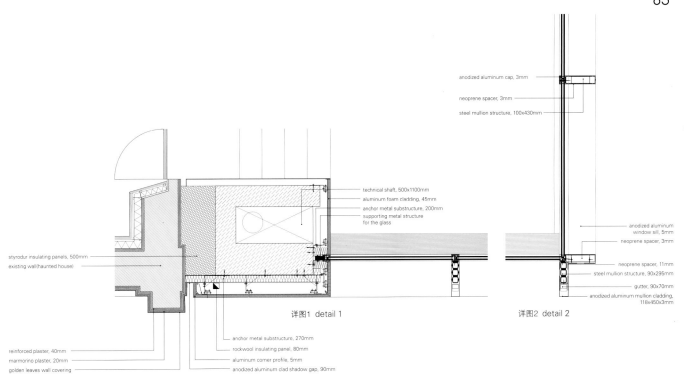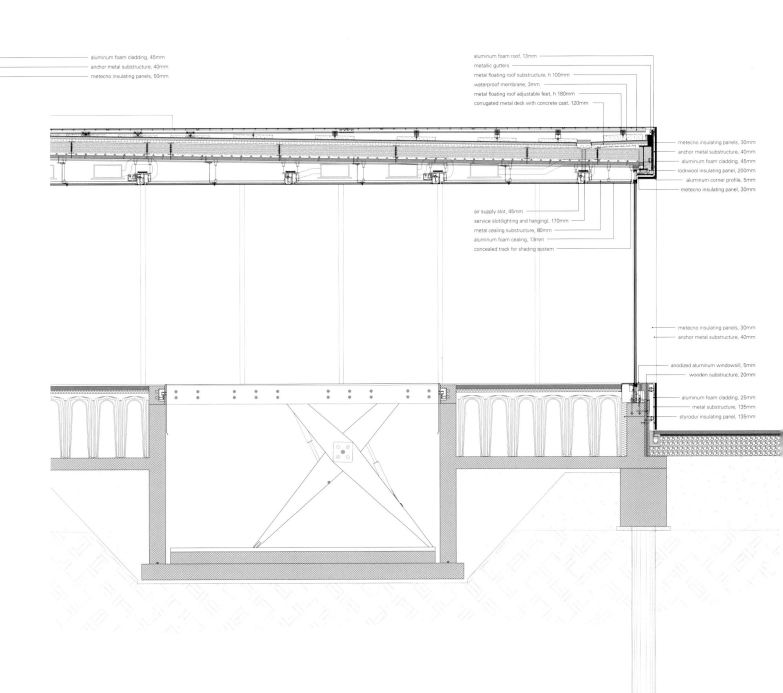

the Fondation's collection is arranged in a hybrid of strict storage and partial display, creating "chambers" where work such as a fleet of artists' cars can be unpacked or half opened to the public.

The freestanding object to the east of the Great Hall, dubbed the Cisterna, is divided in three rooms with three interior "pulpits" connected to an exterior balcony. Its configuration suggests a precise industrial need that now reads as a quasi-religious environment.

The Cinema acts as an autonomous cell within the compound. With large bi-fold doors, it can be instantly connected to the courtyard. Inside, the raked seating can be converted into a flat floor, allowing the space to be used for staging outdoor events or as additional, covered gallery space.

Four "houses" that face the courtyard to the north and an abandoned garden to the south accommodate Fondazione offices and permanent galleries. Within their confines sits the "Haunted House", an existing building with its exterior covered entirely in gold leaf. Inside, the intimate scale of its interiors generates a "domestic" setting for specific works.

Adjacent, the Podium forms the center of the compound, sitting at the intersection of the two perpendicular axes through the site. This addition combines two volumes of very different qualities: a fully glazed, column-free podium on the ground floor. Resting on top is another gallery space clad in aluminum foam, with a bubbled pattern. Both galleries provide large, multi-purpose areas for temporary exhibitions and events.

Another major addition to the complex is a nine-story tower in the north-west corner of the compound. The tower houses site specific installations, as well as a restaurant, a reception space and guest facilities. The other floors, with gradually increasing heights, will accommodate site-specific installations and provide alternating views of the compound and the city.

项目名称：Fondazione Prada / 地点：Milan, Italy
建筑师：OMA / 负责合伙人：Rem Koolhaas, Chris van Duijn / 项目指导：Federico Pompignoli
本地建筑师：Massimo Alvisi / 行政管理建筑师：Alvisi Kirimoto & Partners, Atelier Verticale
结构工程师：Favero&Milan / 机电工程师：Favero & Milan, Prisma Engineering
造价咨询师：GAD / 声学工程师：Level Acoustics / 布景设计：Ducks Sceno / 防火工程师：GAE Engineering
初步设计：Sam Aitkenhead, Doug Allard, Andrea Bertassi, Aleksandr Bierig, Eva Dietrich, Paul-Emmanuel Lambert, Jonah Gamblin, Stephen Hodgson, Takuya Hosokai, Jan Kroman, Jedidiah Lau, Francesco Marullo, Vincent McIlduff, Alexander Menke, Aoibheann Ni Mhearain, Sophie van Noten, Rocio Paz Chavez, Jan Pawlik, Christopher Parlato, Ippolito Pestellini Laparelli, Dirk Peters, Andrea Sollazzo, Michaela Tonus, Jussi Vuori, Luca Vigliero, Mei-Lun Xue
最终设计：Anna Dzierzon, Jonah Gamblin, Hans Hammink, Ross Harrison, Matthew Jull, Vincent Konate, Taiga Koponen, Vincent McIlduff, Andres Mendoza, Susanan Mondejar, Sasha Smolin, Michaela Tonus
施工图设计：Katarina Barunica, Marco Cimenti, Cecilia Del Pozo Rios, Anita Ernodi, Felix Fassbinder, Peter Feldmann, Siliang Fu, Jonah Gamblin, Romina Grillo, Clive Hennessey, Taiga Koponen, Roy Lin, Debora Mateo, Vincent McIlduff, Andres Mendoza, Arminas Sadzevicius, Magdalena Stanescu, Lingxiao Zhang
施工管理：Mateo Budel, Marco Cimenti, Andrea Giovenzana, Nicolas Lee, Victor Pricop, Pawel Panfiluk, Caterina Pedo, Stefano Tagliacarne, Luigi Fumagalli, Andrea Vergani, Nicola Panzeri, Simone Bart
客户：Fondazione Prada
用地面积：13,575m² / 建筑面积：7,780m² / 总建筑面积：22,950m² / 总公共面积：13,670m² / 总私人面积：9,280m²
设计时间：2008 / 施工时间：2014至今 / 竣工时间：2015(phase 1), 2016(phase 2)
摄影师：©Attilio Maranzano (courtesy of Fondazione Prada) - p.60~61
©Bas Princen (courtesy of Fondazione Prada) - p.48~49, p.50~51, p.54~55, p.56~57, p.58, p.64~65, p.66, p.67

作为新型产业的艺术 | Art as the New Industry

卡托维兹西里西亚博物馆
Riegler Riewe Architekten

　　卡托维兹市的历史与重工业和矿业密切相关,为城市留下了独特的人造景观,大型工业厂房和建筑,在人们的集体意识中成为公认的文化遗产,赋予这座城市一种身份。博物馆坐落在前身为"华沙矿区"的地方,毗邻市中心。

　　本设计运用了从外观上看几乎毫无痕迹的干预手段,理念是建造一座可提供各种服务的昂贵的博物馆。

　　为了向原矿区致敬,继承和发挥场地的原有功能,大型的功能空间全部被安排在地下。从外部只能看到抽象的玻璃立方体,这些玻璃立方体与旧的历史建筑和谐地融为一体,为地面下方的展区提供日光照射。其中一座玻璃立方体建筑设有行政、开发和气象控制功能区,新的通道,广场和绿化带网络构成了一个高端的公众休闲场所。

　　通过新建的电梯,参观者可以到达原来矿区起重机架的顶端,俯瞰整个卡托维兹市的景色。

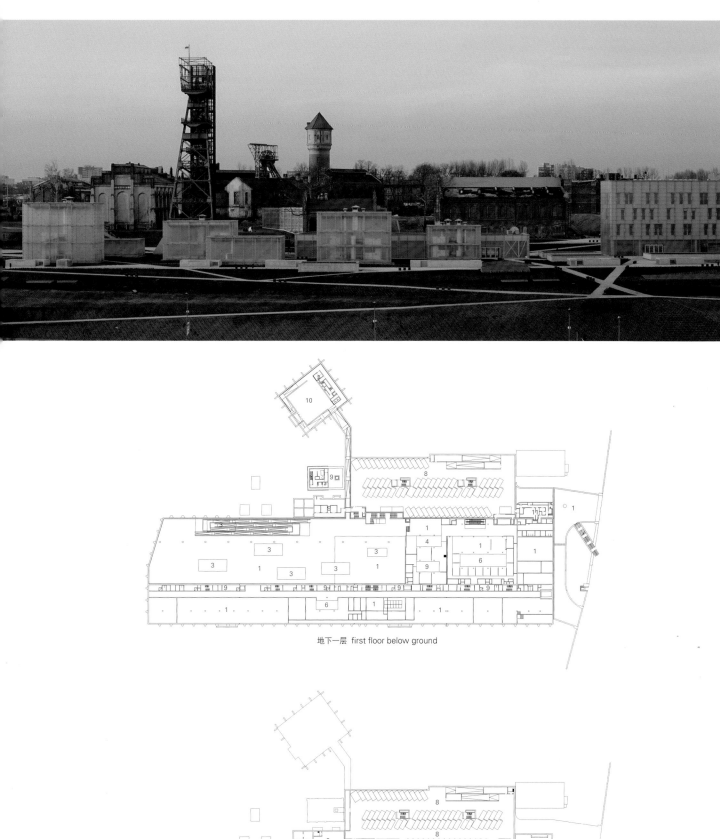

地下一层 first floor below ground

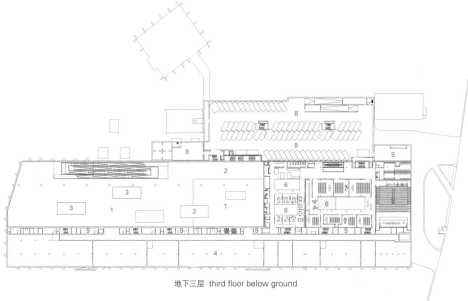

地下三层 third floor below ground

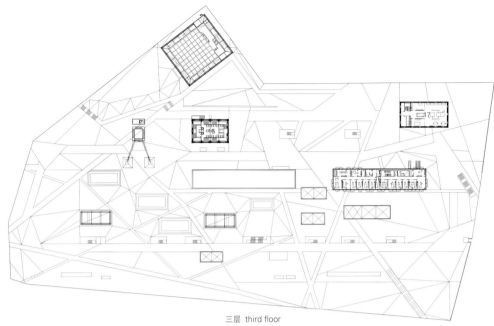

三层 third floor

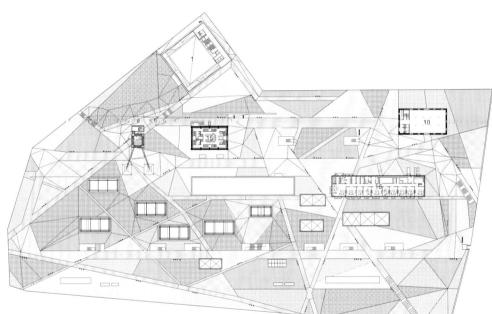

二层 second floor

1 空气间层
2 门厅展区
3 玻璃盒
4 储藏室
5 礼堂
6 中庭
7 会议室 & 讲室
8 地下停车场
9 建筑设备
10 临时展厅
11 厨房
12 入口门厅展区
13 入口门厅会议室/讲室/办公室
14 员工食堂
15 餐厅/酒吧
16 行政办公室
17 工作坊/工作室

1. air space
2. foyer exhibition space
3. glass box
4. storage
5. auditorium
6. patio
7. conference & seminar room
8. underground parking
9. building services
10. temporary exhibition
11. kitchen
12. entrance foyer exhibition area
13. entrance foyer conference/ seminal/administration
14. staff cafeteria
15. restaurant/bar
16. administration
17. work shop/studio

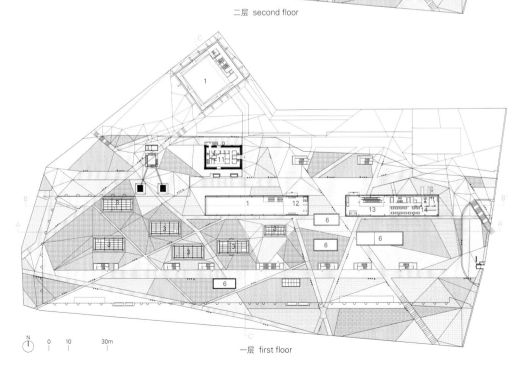

一层 first floor

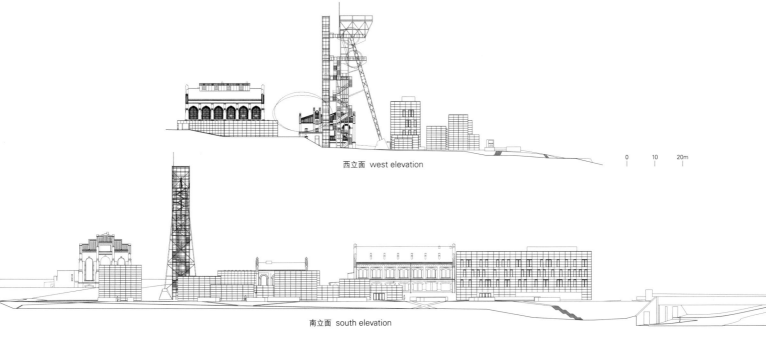

西立面 west elevation

南立面 south elevation

东立面 east elevation

北立面 north elevation

1. aluminium clamping profile
2. 2×10mm toughened safety glass with textured surface
3. 80/56mm steel T-profile supporting structure
4. servomotor
5. 10mm toughened glass with textured surface
6. thermal insulation panel: mineral wool 30mm +140mm covered with aluminium sheet; galvanised steel panel 1mm; plasterboard 12.5mm
7. aluminium shutters, mechanically steered
8. post-and-beam facade structure with system profiles, coated
9. facade ventilation unit
10. fixed glazing: 2×5mm laminated safety glass +16mm cavity +8mm toughened glass
11. floor construction: rubber floor 5mm; screed 55mm; polythene sheet, two layers; impact sound insulation 20mm; lightweight concrete 70mm; 250mm reinforced concrete; plaster

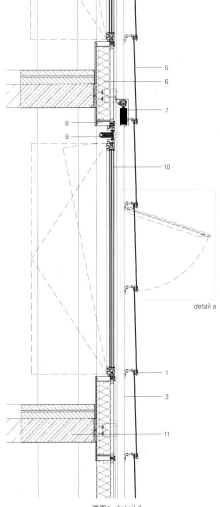

详图1 detail 1

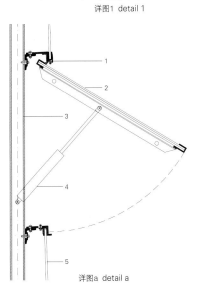 

详图a detail a

Silesian Museum in Katowice

The history of the Katowice city is closely tied with the themes of heavy industry and mining. They have left behind distinctive artificial landscapes, industrial complexes and buildings, and are anchored in the collective consciousness as an unmistakable cultural heritage endowing a sense of identity. The museum is located on the premises of the former "Warszawa mine", directly adjacent to the city center.

The design uses interventions that are nearly imperceptible from the outside and is based on the idea of creating an expansive museum with diverse offerings.

Borrowing from, and as homage to the former function of the terrain, the spatial program was placed entirely below ground. Only the abstract glass cubes, which provide daylight for the exhibition levels below – one of them houses administration, development, climate control – are visible from the outside and meld harmoniously in the ensemble of existing historical structures. The new network of paths, squares, and green areas give rise to a gracefully built public recreational area.

Through the addition of a lift, visitors can access the existing hoist frame and obtain a view over all of Katowice.

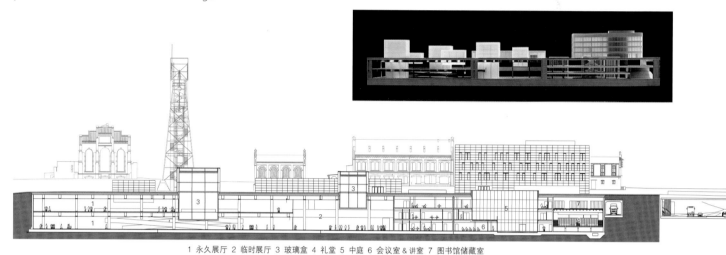

1 永久展厅 2 临时展厅 3 玻璃盒 4 礼堂 5 中庭 6 会议室&讲室 7 图书馆储藏室
1. permanent exhibition 2. temporary exhibition 3. glass box 4. auditorium 5. patio 6. conference & seminar room 7. library storage
A-A' 剖面图　section A-A'

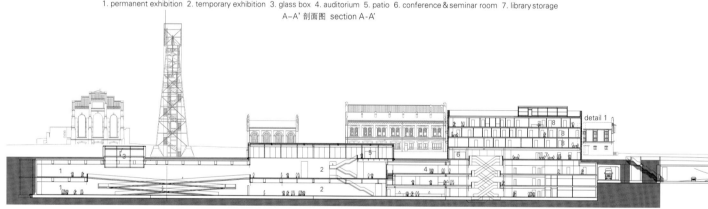

1 永久展厅 2 门厅展区 3 玻璃盒 4 商店(博物馆) 5 入口门厅展区 6 入口门厅会议室/讲室/办公室 7 员工食堂 8 办公室
1. permanent exhibition 2. foyer exhibition space 3. glass box 4. shop(museum) 5. entrance foyer exhibiton area
6. entrance foyer conference/seminal/administration 7. staff cafeteria 8. administration
B-B' 剖面图　section B-B'

1 临时展厅 2 门厅展区 3 储藏室
1. temporary exhibition 2. foyer exhibition space 3. storage
C-C' 剖面图　section C-C'

项目名称：Schlesisches Museum Katowice / 地点：ul. Tadeusza Dobrowolskiego 1, Katowice, Poland / 建筑师：Riegler Riewe Architekten ZT-Ges.m.b.
项目经理：Paulina Kostyra-Dzierżęga / 助理：Anna Zbieranek, Markus Probst, Nicole Lam, Mikołaj Szubert-Tecl, Lavinia Floricel, Minoru Suzuki, Bettina Tòth, Bartłomiej Grzanka, Tomasz Kabelis-Szostakowski, Dorota ŻZurek, Paweł Skóra / 建筑承包商：Budimex S.A. / 结构设计咨询：Firma Inżynierska Statyk
客户：Muzeum Śląskie / 用地面积：27,332m² / 建筑面积：3,343m² / 总建筑面积：25,067m² / 容积：228,702m³ / 造价：EUR 65.1 milion
设计时间：2007 / 施工时间：2011 / 竣工时间：2013
摄影师：©Paolo Rosselli (courtesy of the architect)-p.68~69, p.70, p.79^{bottom}, p.80, p.81 / ©Wojciech Kryński (courtesy of the architect)-p.72~73, p.74~75, p.76, p.77, p.79^{top}

敦刻尔克地区当代艺术基金会
Lacaton & Vassal

作为新型产业的艺术 | Art as the New Industry

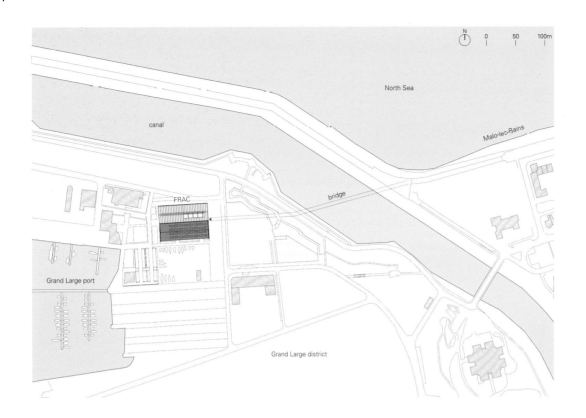

FRAC是一座用于收藏当地的当代艺术收藏品的建筑。

这些收藏品通过现场展示以及出借给画廊和博物馆的形式进行保存、归档和对公众开放。FRAC的北区位于敦刻尔克港一个被称作Halle AP2的旧船仓库中。Halle AP2是一座标志性的单体建筑,内部空间超大,光线充足明亮,令人印象深刻,使用潜力非凡。

项目设计的基本理念是使FRAC成为这一新区域进一步发展的催化剂,同时能够整体上保留原有的Halle AP2风貌。为了实现上述理念,设计师保留了原先的建筑,又设计了一座新的建筑,紧邻原先的建筑,大小完全一致,面向大海,并包含了FRAC的功能。

新FRAC大楼精致巧妙地与原来的仓库并置,既不争奇斗艳,也不黯然失色。这种复制是对原建筑的认同;新建筑外围护结构轻盈灵动,符合生物气候学原理,高效的预制结构决定了内部平台的自由、灵活和富于变化的特性,少有空间上的限制约束,符合项目需求。

透明的建筑表皮可以使参观者从背后看到用来储藏艺术品的不透明建筑体量。穿过建筑的公共人行天桥(原设计是沿着建筑物的外立面设置的)变成进入Halle AP2和FRAC内部立面的室内一条街。Halle AP2的空间没有固定永久占用,完全处于可以随时利用的状态,既可以和FRAC一起承担各种活动(特殊的临时性展览、大规模艺术作品的创作、特别的操作处理),也可以独立承办一些公共活动(音乐会、商品交易会、表演、马戏表演、体育赛事),充满无限利用可能。

这两座建筑中的任何一座都可以保持独立,也可以合而为一。建筑的架构及其现在的质量只能进行尽可能少的、有针对性的且有限的干预措施。在预算范围内,FRAC大楼顺利建成,Halle AP2的环境和公共设施均得以完善,这都多亏了该项目的最优化设计。

因此,该项目创造了一套庞大的公共资源:其灵活的空间可以用来举办从日常展览到大型艺术盛宴等多种规模的活动;不仅引起本地区的关注,也引起了欧洲乃至全世界的关注,强化了敦刻尔克港口的重建工作。

FRAC Dunkerque

The FRAC houses regionally assembled public collections of contemporary art.

These collections are conserved, archived and presented to the public through on-site exhibitions and by loans to both galleries and museums. The North region of FRAC is located on the site of Dunkerque port in an old boat warehouse called Halle AP2. The Halle AP2 is a singular and symbolic object. Its internal volume is immense, bright, impressive. Its potential for uses is exceptional.

To implant the FRAC, as a catalyst for the new area, and also to keep the Halle in its entirety become the basic idea of our project. To achieve this concept, the project creates a double of the Halle, of the same dimension, attached to the existing building, on the side which faces the sea, and which contains the program of the FRAC.

The new building juxtaposes delicately without competing nor fading. The duplication is the attentive response to the identity of the Halle; Under a light and bioclimatic envelope, a prefabricated and efficient structure determines free, flexible and evolutionary platforms, with few constraints, fit to the needs of the program.

The transparency of the skin allows to see the background vision of the opaque volume of the artworks reserves. The public

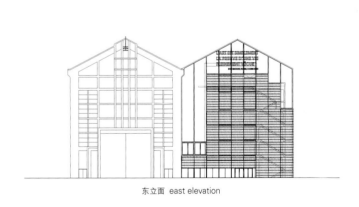

东立面 east elevation

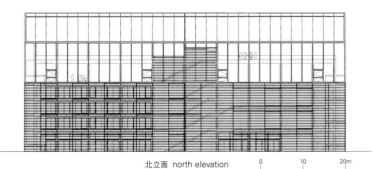

西立面 west elevation

北立面 north elevation

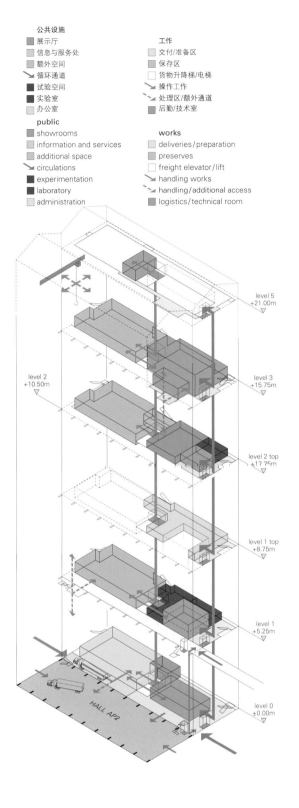

2012.05.10

2012.09.13

2013.06.06

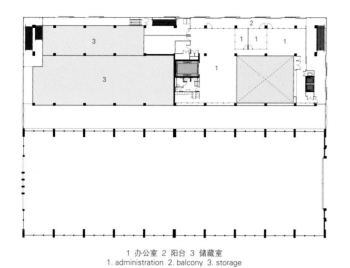

1 办公室 2 阳台 3 储藏室
1. administration 2. balcony 3. storage
三层 second floor

1 瞭望台 2 办公室食物筹备处 3 会议室
1. belvedere 2. office caterer 3. meeting room
六层 fifth floor

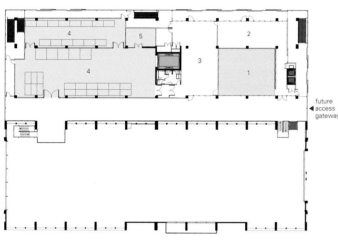
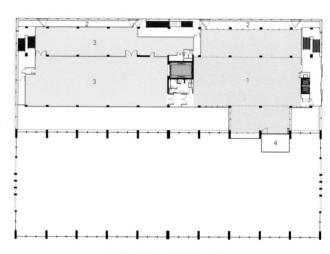

1 展室 2 休息室 3 实验工作室
4 储藏室 5 试听设备储藏室 6 平台
1. exhibition room 2. lounge 3. experimentation atelier
4. storage 5. audio & visual storage 6. platform
二层 first floor

1 论坛 2 阳台 3 储藏室 4 平台
1. forum 2. balcony 3. storage 4. platform
五层 fourth floor

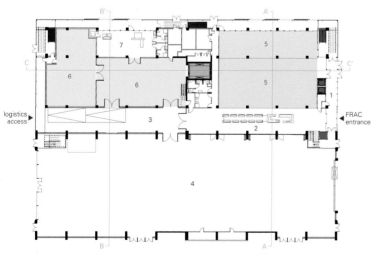
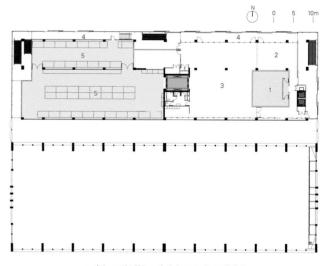

1 入口大厅 2 咖啡厅/FRAC接待处 3 装卸货处
4 Halle AP2 5 展室 6 交通运输区域 7 工作室
1. entrance hall 2. cafe/FRAC reception 3. loading dock
4. Halle AP2 5. exhibition room 6. transport area 7. atelier
一层 ground floor

1 暗室 2 陈列柜 3 实验室 4 阳台 5 储藏室
1. dark room 2. showcase 3. laboratory 4. balcony 5. storage
四层 third floor

footbridge (previously planned along the facade) which crosses the building becomes a covered street entering the Halle and the internal facade of the FRAC. The Halle AP2 will remain a completely available space, which can work either with the FRAC, in extension of its activities (exceptional temporary exhibitions, creation of large scale works and particular handlings) or independently welcome public events (concert, fairs, shows, circus, sport) and which enriches the possibilities of the area. The functioning of each of the buildings is separated, or combined. The architecture of the Halle AP2 and its current quality make sufficient minimal, targeted and limited interventions. Thanks to the optimization of the project, the budget allows the realization of the FRAC and the setting up of conditions and equipment for public use of the Halle AP2.

The project so creates an ambitious public resource, of flexible capacity, which allows work at several scales from everyday exhibitions to large-scale artistic events, of regional but also European and international resonance, which consolidates the redevelopment of the port of Dunkerque.

项目名称：FRAC (Regional Contemporary Artwork Collection) of the North Region
地点：Dunkerque, France
建筑师：Anne Lacaton, Jean Philippe Vassal
合作方：Florian de Pous, chief project, Camille Gravellier
施工监理：Yuko Ohashi
结构/机械系统：Secotrap / 金属结构：CESMA
造价咨询：Vincent Pourtau / 防火安全咨询：Vulcanéo
功能：artwork reserves, exhibition rooms, education
造价：EUR 12M
面积：9,157m² _ new building, 1,972m² _ existing hall
设计时间：2010 / 施工时间：2011 / 竣工时间：2013
摄影师：©Philippe Ruault - p.82~83, p.85bottom, 88~85, p.91, p.92~93
©Communauté Urbaine de Dunkerque(courtesy of the architect) - p.84, p.85top

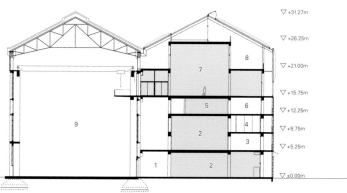

1 咖啡厅/FRAC接待处 2 展室 3 休息室 4 办公室
5 暗室 6 陈列柜 7 论坛 8 瞭望台 9 Halle AP2
1. cafe/FRAC reception 2. exhibition room 3. lounge 4. administration
5. dark room 6. showcase 7. forum 8. belvedere 9. Halle AP2
A-A' 剖面图 section A-A'

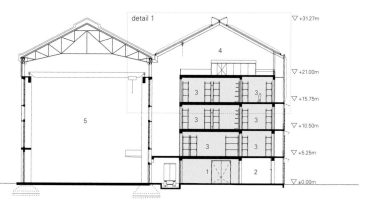

1 交通运输区域 2 工作室 3 储藏室 4 瞭望台 5 Halle AP2
1. transport area 2. atelier 3. storage 4. belvedere 5. Halle AP2
B-B' 剖面图 section B-B'

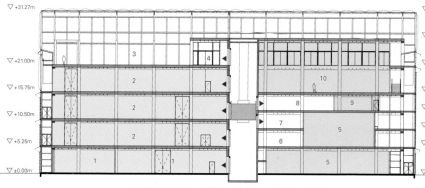

1 交通运输区域 2 储藏室 3 瞭望台 4 会议室 5 展室
6 工作室 7 办公室 8 实验室 9 暗室 10 论坛
1. transport area 2. storage 3. belvedere 4. meeting room 5. exhibition room
6. atelier 7. administration 8. laboratory 9. dark room 10. forum
C-C' 剖面图 section C-C'

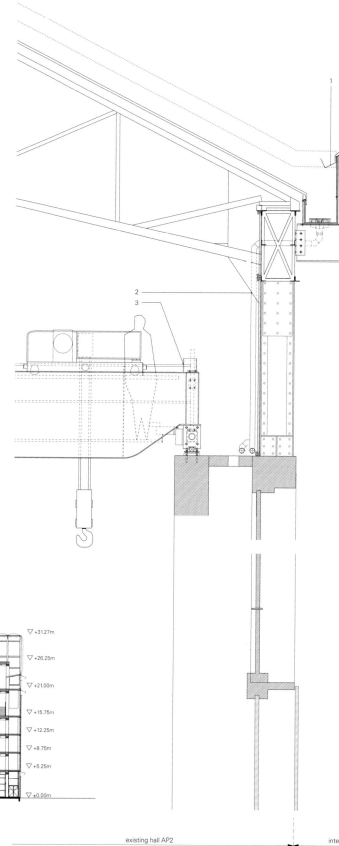

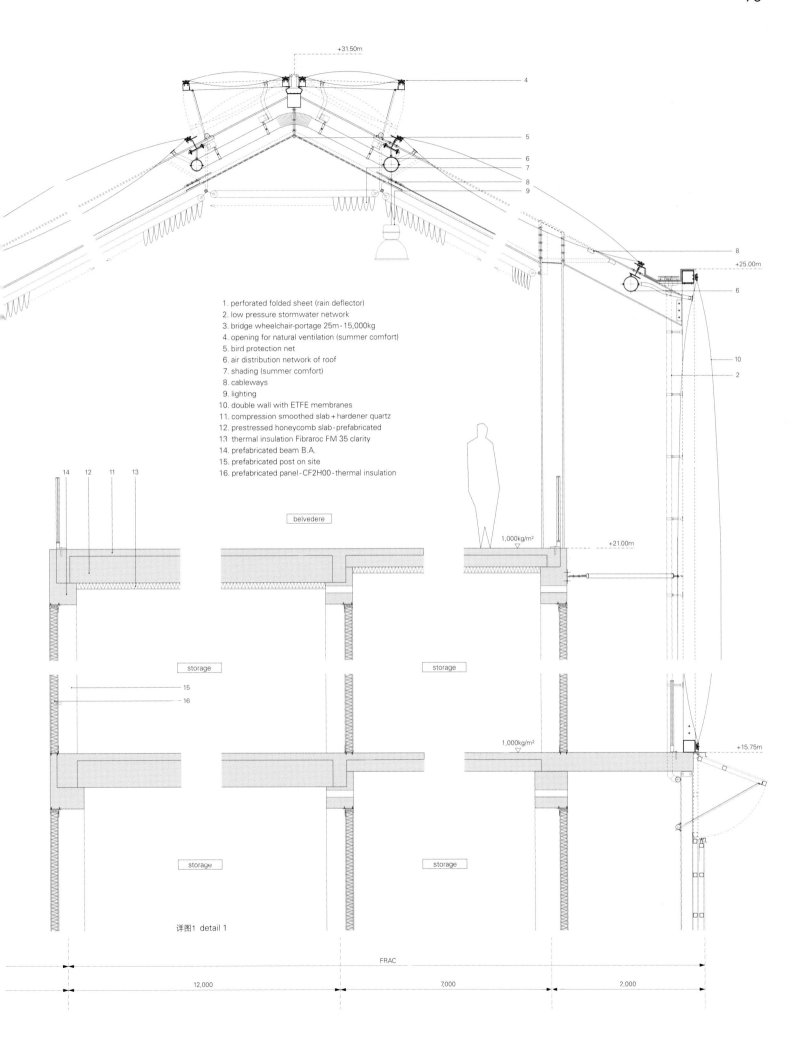

详图1 detail 1

家居生态

景观设计的三种方式

在不涉及设计师的情况下，建筑值得称赞的一方面就是其与周围景致的交相辉映。对J.B.杰克逊或伯纳德·鲁道夫斯基而言，新墨西哥的Taos Pueblos或安达卢西亚的Pueblos Blancos（白色城镇）的一致性可以说创造了第二种自然。建筑环境成为自然景观的一部分。这种休戚与共一方面来自于数个世纪以来人们合理结合气候、地质情况与地形进行空间实践的宝贵经验，另一方面则是由社会与文化因素所引起的。

尽管这第二种自然似乎自发出现在乡土建筑传统中，但如今在"有经验的"建筑中也被广泛应用。比如说，有时候山地建筑和所谓的有机建筑的建造，这是对自然特征可笑的模仿。在其他时候，建筑师有意识地试图将自然特征融入到建筑形式中，例如，墨西哥的迪奥狄华肯金字塔或者埃及的吉萨金字塔。然而，随着工业革命所带来的科技发展，挑战自然变成了更为普遍的动机，而非与大自然产生共鸣。回顾上个世纪，继一部分人对技术盲目崇拜而产生了骄傲自大的态度之后，符合自然特点的设计很快浮出了水面。这种与自然环境相结合的设计方式不但表达了对自然的尊敬，也寻求与大自然的和谐共存。现如今，虽然可以利用丰富的建筑材料和各种技术来突破地势、重力、气候等因素的限制，但是设计师的创造力仍然受到将艺术与景观融为一体的决心的挑战，即目前流行的创造新型家居生态的决心。

A common facet in the praise on the architecture without architects was its capacity to mingle with the surrounding environment. For J.B. Jackson or for Bernard Rudofsky, the uniformity of the Taos Pueblos in New Mexico or the Pueblos Blancos (The White Towns) of Andalucía created a second nature, as it were. The built environment was one with the natural landscape. On the one hand this symbiosis was accomplished by centuries of sensible adaptation of spatial practices to climate, geology and topography. On the other hand, it was induced by social and cultural factors.

While this second nature seemingly surfaced spontaneously in vernacular building traditions, it has also been relentlessly pursued in "learned" architecture. Sometimes as a derisory imitation of natural features, creating, for example, mountain buildings and so-called organic constructions. At other times, architects consciously attempted to translate natural features into built-form erecting, for example, the pyramids of Teotihuacan in Mexico or Giza in Egypt. However, with the technological developments brought forth by the Industrial Revolution, challenging nature became a more pervasive motivation than resonating with it. Over the last century, next to the arrogance of some technological fetishism a parallel drive to design along with nature has surfaced. The latter was often characterized by an approach to the natural environment that combined a sympathetic reverence to it while searching a meaningful coexistence with it. Nowadays, notwithstanding the abundant material and technological options available to transcend the constraints imposed by topography, gravity, climate, and so on, the creative energy of designers is still challenged by a will to conflate art with the landscape, a pervasive determination to create new ecologies of domesticity.

Dwell How

Ecologies of Dom

Three Ways of Designing the Landscape

永恒之屋 / Alberto Campo Baeza
蒙桑双宅 / João Paulo Loureiro
卡特彼勒住宅 / Sebastián Irarrázaval
八岳山住宅 / Kidosaki Architects Studio
MO住宅 / Gonzalo Mardones Viviani
"秋之屋" / Fougeron Architecture
向日葵住宅 / Cadaval & Solà-Morales
桧原住宅 / SUPPOSE DESIGN OFFICE

家居生态：景观设计的三种方式
/ Nelson Mota

House of the Infinite / Alberto Campo Baeza
Two Houses in Monção / João Paulo Loureiro
Caterpillar House / Sebastián Irarrázaval
House in Yatsugatake / Kidosaki Architects Studio
MO House / Gonzalo Mardones Viviani
Fall House / Fougeron Architecture
Sunflower House / Cadaval & Solà-Morales
House in Hibaru / SUPPOSE DESIGN OFFICE

Ecologies of Domesticity: Three Ways of Designing the Landscape
/ Nelson Mota

度假屋，以及富裕人士用于避开现代生活的繁琐的、风景如画的休闲别墅，都是建筑媒体最喜欢报道的主题之一。的确，哪个建筑师不曾见过像弗兰克·劳埃德·赖特的流水别墅、密斯·凡·德·罗的范斯沃斯住宅、奥斯卡·尼迈耶的独木舟之家、胡安·安东尼奥·科德奇的乌尔加德别墅或阿达尔贝托·利贝拉的马拉帕特别墅这样壮观的图景？这些经典案例都使人们想起如梦似幻的氛围。在那样的氛围中，人工建筑与大自然和谐相融，创造出新的地势，并将自然景色引入家中。诚然，马拉帕特别墅的案例证明了人们具有创作这种艺术与大自然水乳交融的作品的潜力以及由此带来的诱惑。

库尔乔·马拉帕特在他的自传体小说La Pelle（《皮肤》）中写道，纳粹陆军元帅埃尔温·隆美尔曾经拜访过他在卡普里岛的房子。隆美尔痴迷于这"世界上最美的景色"，问马拉帕特，这所房子在他住进去之前就存在还是他自己设计并建造了这所房子。马拉帕特的回答让人非常震惊。他告诉隆美尔这所房子早就存在，但是，是他"设计了风景"。1980年，美国建筑师约翰·海杜克应意大利的《Domus》杂志之邀为马拉帕特别墅撰写评论，他将这座住宅描述为"一所充满宗教仪式与礼制的房子"。事实上，海杜克宣称："这是一所神秘的房子，能立刻在过去祭品的头或角上带来爱琴海的寒意，是在意大利上演的一部古老剧本。"²

正如海杜克所说，这所"神秘的房子"的神秘感已经被法国电影制作人让·吕克·戈达尔抢先捕捉到，并运用于1963年他拍的电影《蔑视》中。在该电影里，戈达尔运用马拉帕特别墅房子与自然之间模糊

The holiday house and the well-to-do's scenic retreat from the nitty-gritty affairs of modern life are one of the favourite subjects of architectural media. Indeed, which architect hasn't been confronted with the splendours images of the likes of Frank Lloyd Wright's Fallingwater, Mies van der Rohe's Fransworth House, Oscar Niemayer's Casa das Canoas, Juan António Coderch's Casa Ugalde, or Adalberto Libera's Casa Malaparte? These classic examples suggest dreamlike atmospheres where the built artefact flirts with nature to build new topographies and to domesticate the landscape. To be sure, the case of Casa Malaparte testifies to the creative potential of, and the allurement caused by, this intercourse between art and nature. In his autobiographic novel La Pelle (The Skin), Curzio Malaparte reports a visit to his house in Capri by the Nazi Marshal Erwin Rommel. Bewildered by "the most beautiful scenery in the world" Rommel asks Malaparte whether the house was already built before he moved in or whether he had designed and built it himself. Malaparte's reaction was quite striking. He answered Rommel that yes, the house already existed, but he had "designed the scenery".¹ When, in 1980, the American architect John Hejduk was invited to write a review of Casa Malaparte for the Italian journal Domus, he described it as "a house of rituals and rites". Indeed, Hejduk asserted, "it is a house of mysteries, it at once brings forth the chill of the Aegean on the horn head of past sacrifices, and it is an ancient play placed in an Italian light."²

The enigmatic qualities of this "house of mysteries", as Hejduk put it, had been previously caught by the lens of French filmmaker Jean-Luc Godard in his movie Le Mépris (Contempt), produced in 1963. Godard uses the blurred distinction between house and nature in Casa Malaparte to create the background against which the two protagonists of the movie, Paul (Michel Piccoli) and Camille (Brigitte Bardot), discover their own boundaries and thresholds. Indeed, as a reviewer notes, "the characters trace the thresholds of the roof and call into the landscape, only to be confronted with their own boundaries.

新墨西哥的Taos Pueblos村
Taos Pueblos in New Mexico

安达卢西亚的Pueblos Blancos村
Pueblos Blancos of Andalucía

墨西哥的迪奥狄华肯金字塔
The Pyramids of Teotihuacan in Mexico

的区别创造了背景，使得电影中的两个主人公保罗（米歇尔·皮寇利饰演）和卡米尔（碧姬·芭铎饰演）发现了横亘在他们之间的界线与门槛。实际上，正如一位评论家指出的："人物角色顺着屋顶的边界走入风景之中，仅仅是为了面对他们之间的界线。马拉帕特别墅的建筑形式与空间格局进一步促进了上述情况的发生，在许多方面，这所房子就如同一座寺庙和一座监狱，表现了他们之间的关系。"³

的确，马拉帕特别墅成了一个几乎神话般的实例，展示了人、建筑和自然之间那诗情画意、神秘莫测、神奇魔幻般的共生关系。这种魔力至今依然存在，并且以当仁不让之势出现在前面介绍的建筑项目中。在不同物种生物体之间的种间关系上，我们可以将之分成几种不同类型的共生关系，包括寄生、偏利共生与互利共生。同样，建筑也在自然与人造物体之间创造了不同的关系，包括引人注目的大胆尝试、友好共存与纯粹的合作关系。

引人注目的大胆尝试

卡特彼勒住宅与周围的景色展现了人工与自然的巧妙结合。人工指的是形成圣地亚哥大都会景色的那些东西，而自然指的是神话般的安第斯山脉，绵延在南美的西海岸线上。卡特彼勒住宅由Sebastián Irarrázaval设计，坐落在智利首都圣地亚哥郊区的巴尼特。Irarrázaval使用航运集装箱进行大胆构建，为这所房子创造出一系列空间。显著的大胆设计定义了各建筑体量的关系，使得该房子顺地势而建。悬臂式体量、旋转模块以及混凝土和耐候钢表面的巧妙结合，都是为了使房屋与周围多岩石的粗糙环境产生共鸣。在野兽派艺术的鼎盛时期，Irarrázaval巧妙地探索了利用航运集装箱作为构建房屋的基础模块这种粗野主义的诗歌，创造出对智利景色诗一般的敬意。

还是在智利，但这一次是在瓦尔帕莱索地区，由冈萨罗·马多内斯·薇薇安娜设计的MO住宅，也是利用景观来塑造建筑的空间结构。

This event is facilitated through the form and space of Casa Malaparte and in many ways the house becomes an expression of their relationship, as a temple and a prison."³
The Casa Malaparte became indeed an almost mythical instance of the poetic, mysterious and magical symbiosis between people, architecture and nature. This fascination persists today and it is compellingly illustrated in the projects featured ahead. In biological interactions between organisms of different species, we can identify different types of symbiosis, ranging from parasitism, to commensalism, to mutualism. By the same token, the projects that follow similarly create different relationships between nature and the artifact, stretching from conspicuous bravura, to amiable coexistence to pure partnership.

Conspicuous Bravura

On the plot where Catapillar House stands the landscape offers a magnificent combination of artifacts and nature. The artifacts are that which form the cosmopolitan scenery of Santiago and the nature of the Andes, the mythical mountain range that stretches all along the western coast of South America. Caterpillar House, was designed by Sebastián Irarrázaval and is located in Lo Barnechea, a commune on the outskirts of Chile's capital city, Santiago. Irarrázaval uses a bold composition of shipping containers to create the sequence of spaces that define the house. Noticeable moments of design bravura define the relation of the volumes that conform the house to the topography. Cantilevered volumes, rotated modules and a careful combination of concrete and weathering steel surfaces are deployed to achieve deliberate resonances with the rocky and harsh surrounding landscape. As in the heyday of Brutalism, Irarrázaval ingeniously explores the rough poetry of the basic module of the house, the shipping container, to create a poetical homage to the Chilean landscape.

Yet in Chile, but this time in the Valparaiso region, the MO House designed by Gonzalo Mardones Viviana also takes advantage of the landscape to shape the spatial configuration of the building. In this case, however, the designer keenly creates multiple filters to explore gradients of light and shadow, and to build up porous thresholds between the concrete volumes and the surrounding scenery. The skylights and patios, the large eaves and beams and the projecting volume block placed on the roof of the main floor are all design features that

意大利卡普里岛的马拉帕特别墅
Casa Malaparte in Capri, Italy

然而,在这个项目里,设计师热衷于设计许多滤光器以探索光与影的渐变,在介于混凝土体量与周边景致之间的门槛上方也做了许多开孔设计。天窗和天井、宽大的屋檐和横梁、主楼层屋顶突出的建筑体量模块,都是这个项目的设计特点,表达了设计师想要融入周边景色的强烈愿望。

在上面提到的位于智利的案例中,建筑的剖面设计都证明了设计师致力于复制地形情况,而由城户崎建筑师工作室设计的日本长野八岳山住宅,却故意挑战项目所在地的地形地貌。房间在斜坡上水平延伸出来,其设计融合并再造了周围景色。住宅的大半部分都悬在空中,悬空的巨大楼板通过两根斜撑钢柱来支撑。这确实是一个重大的设计决定,正如设计师本人所说的,探讨了非传统的结构解决方案:"在建筑作为主导因素凌驾于周围环境之上和与周围自然环境和谐共存之间实现了适当的平衡。"

但是,这种平衡并不是那么容易就能达到的。确实,作为旧金山Fougeron Architecture建筑事务所的负责人,"秋之屋"的设计师安妮·富热隆说:"将房子建在荒野是一种激进的行为。"诚然,位于加州大索尔山,高于太平洋海平面76m的地方,"秋之屋"显示了建筑与自然之间的矛盾关系。一方面,将房子解构成几部分,形成高低错落的楼层和天花板,所看到的视野和感受到的氛围也变化万千,令人不得不更加敬佩风景的力量。另一方面,房屋的南立面与顶部都由铜箔覆盖,用堆(卵)石混凝土做地基,还有悬挑的体量,这些都表明"秋之屋"的设计非常大胆。此大胆设计,正如设计师所说的,"是让自然环境去适应房屋的选址"。

友好共存

在上面提到的案例中,建筑师有共同的驱动力,就是在建房子的地方探索一种创造性的张力。而在这个部分,我将会探讨一些案例,在

express a strong will to tame the landscape.

While in the Chilean cases discussed above, the section of the building testifies to the designers' commitment to replicate the topographical situation, the House in Yatsugatake, designed by Kidosaki Architects Studio, shows a deliberate drive to challenge the features of the site. The house expands horizontally over the slope, incorporating and reframing the scenery into the design. A large section of the house detaches from the ground, creating an overhanging structure supported only by two diagonal steel columns. This is indeed a fundamental design decision that explores unconventional structural solutions to achieve, as the architect asserts, "the proper balance between a dominance over and a harmony with the surrounding natural environment."

This balance is not always easy to achieve, though. Indeed, "placing form on wilderness is a radical act", says the designer of Fall House, Anne Fougeron, the principal of San Francisco based Fougeron Architecture. To be sure, perched 76 meters above the Pacific Ocean in California's mountainous Big Sur, Fall House suggests an ambivalent relation between architecture and nature. On the one hand, the deconstruction of the house in several volumes, creating multiple level changes in the floors and ceilings and a wide array of perspectives and atmospheres pays tribute to the power of the landscape. On the other hand, the copper cladding that wraps the southern facade of the house and its roof, the concrete boulder locking the house to the land, and the cantilevered volumes are clear instances of design bravura that, as the designer admits, "defies natural forms to accommodate the siting."

Amiable Coexistence

In the cases discussed above there was a common drive to explore a creative tension with the situation in which the houses were built. In this section, I will discuss cases where there is a thoughtful coexistence between the built artifact and the natural setting.

The House in Hibaru projected by SUPPOSE DESIGN OFFICE delivers a good example of an amiable coexistence between built form and nature. The main feature of the project is a pitched roof that follows the slope of the plot where the house is built, a stretch of land between a road and a water reservoir in the Japanese city of Fukuoka. Underneath the roof, the designers accommodate the domestic program in split-levels that rep-

智利萨帕利亚尔的MO住宅
MO House in Zapallar, Chile

日本长野的八岳山住宅
House in Yatsugatake in Nagano, Japan

这些案例中，建筑与自然环境和谐共存。

由日本SUPPOSE DESIGN OFFICE建筑事务所设计的桧原住宅是体现建筑与自然和谐共存的典型案例。在日本福冈的一条公路与水库之间的一块地上，倾斜的屋顶随斜坡的走势而建，成为该建筑的主要特点。在屋顶下，设计师也同样顺应地势将房屋内的不同功能区域分布在不同水平面上。房子与水库和周围环境的关系经过精心设计，釉面砖铺设的小路犹如一条丝带从三面环绕着房子，将三者连为一体。这座房子展现了暴露与隐藏、隐私与开放的巧妙设计。

向日葵住宅位于地中海沿岸地处加泰罗尼亚布拉瓦海岸的德拉塞尔瓦海港渔村，由Cadaval和Solà-Morales共同设计。很显然，该设计对房屋融入周围的景色同样非常关注。这座房子被拆分成许多小的单元，能看到更多景色，拥有多样化捕捉阳光与自然光的途径。与此同时，不连贯的外立面可以保护房屋免受大自然的侵害，特别是可以抵御来自于特拉蒙塔那山脉强劲的风对其带来的不良

影响。该项目的设计者声称，"该住宅构建了一种多样化的独特视野"。尽管这样，这仍然是一处神秘的小单元空间。实际上，房屋内部的空间结构并未碎片化，可欣赏到全景的室内空间体验是完整的。

纯粹的合作关系

在以上谈论的案例中，设计师通过利用引人注目的材料和具有表现力的建筑形式，大胆设计并寻求自然与人工之间的友好共存。在这个部分，我将列举两个建筑，它们要表达的是向自然景观默默致敬以及自然环境与建筑形式之间精心设计的伙伴关系。

葡萄牙建筑师João Paulo Loureiro设计的蒙桑双宅体现了技术的大胆创作与景观合理融合之间的微妙平衡。这两栋住宅坐落在葡萄牙米尼奥河岸，靠近蒙桑村，由一块预应力混凝土板连成一个独特的建筑单体。这块预应力混凝土板长84m，与河岸平行。诚然，这个混凝土板就是该建筑与众不同的因素，它既可以是屋顶，下面是走廊，也可以

licate the topography of the site. The relation with the water body and the surrounding context is carefully crafted through a ribbon of glazed surfaces that wrap the building around three sides. This house shows a carefully devised play of concealment and exposition, privacy and openness.

The same concern with a precise framing of the surrounding landscape is noticeable in the Sunflower House designed by Cadaval & Solà-Morales for a site bordering the Mediterranean Sea in El Port de la Selva, a seaside village in the Catalonian Costa Brava. The house is fragmented in multiple volumes that expand the contact with the scenery, creating multiple passages to capture sun and natural light. At the same time, the broken facade offers protection against nature, especially against the strong Tramuntana winds. The authors of the project claim, "the house frames a multiplicity of different and specific views." This is an elusive fragmentation, though. Indeed, instead of partial and segmented, the spatial experience of the interior of the house is panoramic and whole.

Pure Partnership

In the projects discussed above, the architect's design bravura and the search for an amiable co-existence between artifact and nature were often pursued with conspicuous materials and expressive forms. In this section, I will discuss two projects whose expressiveness is conveyed by silent homages to the landscape, well-crafted partnerships between built form and natural environment.

João Paulo Loureiro's Two Houses in Monção negotiate a delicate balance between technological bravura and sensible integration on the landscape. Located in the Portuguese bank of the river Minho, close to Monção, the two houses are combined into a unique formal gesture through a pre-stressed concrete slab that spans 84 meters through the site, parallel to the riverbank. To be sure, this is the element that chiefly defines the project. It is simultaneously a roof, a porch and a platform pierced by a glazed volume and a voluptuous spiral staircase. Furthermore, the concrete slab is carefully combined with granite masonry walls and ebony and pine interior finishing to create rich domestic atmospheres. The slab shelters and protects while, at the same time, exposes the surrounding elements, the sky and the ground.

This fusion between the tangible and the intangible qualities of a building is a persistent trait in Alberto Campo Baeza's archi-

西班牙加的斯的永恒之屋 House of the Infinite in Cádiz, Spain

是平台,一个玻璃建筑和一个令人惊艳的螺旋楼梯从其中穿过。另外,混凝土板与花岗岩砌筑墙、乌木的和松木的内部装饰巧妙结合,形成了丰富的室内景观。混凝土板在起到遮蔽和保护作用的同时,也将蓝天、大地等周围景色呈现给了房屋的主人。

建筑物有形特质与无形特质之间的融合是阿尔贝托·坎波·巴埃萨设计作品中永恒的主题,总是充满了诗意般的共鸣。他设计的永恒之屋就是一个鲜明的例子。实际上,正如阿达尔贝托·利贝拉设计的马拉帕特别墅一样,坎波·巴埃萨设计的永恒之屋也模糊了地上建筑与天上自然景观的界限,暗示这座巨大的罗马洞石之屋是从沙滩浮现而出的。永恒之屋坐落在西班牙安达卢西亚省加的斯市的海边。正如建筑师所说,永恒之屋创造了一个"石头卫城"。设计师有意避免了世俗的设计方案,而是隐藏技术设备的痕迹,整个体量以及其中的空间都棱角分明,这都是靠技术设备来实现的。然而,设计师旨在创造一个永恒的、无边无际的建筑,从安达卢西亚的海岸线一直延伸到永恒,无边无际,形成自然与艺术之间纯粹的合作关系。

家居生态

上述提到的建筑与建筑界家喻户晓的一些经典设计案例有共同之处,都强调建筑与其周围自然环境进行有意义的交流。尽管如此,还是创造了不同的家居生态。我们可以看到,紧随弗兰克·劳埃德·赖特设计的流水别墅之后,有些建筑在技术上大胆尝试,来改造景观,令其备受瞩目。还有一些建筑则努力实现自然景观与建筑的和谐共存,例如,胡安·安东尼奥·科德奇设计的乌尔加德别墅。最后,第三种是设计师强烈地追求建筑与自然之间纯粹的合作关系,正如马拉帕特别墅那样,消除时间与空间的限制。总结全文,库尔乔·马拉帕特与隆美尔的对话犹在耳畔,在这三种家居生态中,设计师不仅仅在设计房子,也在寻求以独特的方式来设计风景。

tecture, always charged with poetic resonances. His VT House is a compelling example of this. Indeed, as Adalberto Libera's Casa Malaparte, also Campo Baeza's VT House blurs the limit between earthly architecture and heavenly natural idylls, suggesting the image of a travertine monolith emerging from the sand. Located on the seaside of Cádiz, a city in the Spanish province of Andalucía, House of Infinite creates "an acropolis in stone", as the architect suggests. There is a deliberate attempt to avoid worldly references, to conceal the technological apparatus that underpin the sharp edges that define the volume and the spaces excavated within it. Rather, the designer's ambition is to create a timeless and boundless artifact, stretching from Andalucía's coast to eternity and to the infinite. A pure partnership between art and nature.

Ecologies of Domesticity

The houses discussed above have in common with the classic examples that are household names to the architecture discipline, a strong drive to create a meaningful exchange with the natural setting on which they are situated. They create different ecologies of domesticity, though. Following the lead of Frank Lloyd Wright's Fallingwater, in some of these cases one can observe a conspicuous use of technological bravura to reshape the landscape. In other cases, prevails an attempt at achieving an amiable coexistence between the natural landscape and the built form, in the same vein as Juan António Coderch's Casa Ugalde, for example. Finally, a third approach can be singled out, in which designers keenly pursue a pure partnership with the landscape, dissolving the physical and temporal limits of the artefact as in Casa Malaparte. Echoing Curzio Malaparte's exchange with Rommel, one could conclude that in these three ecologies of domesticity, more than designing a house, the architects sought distinctive ways of designing the landscape. Nelson Mota

1. Vittorio Savi, "Orphic, Surrealistic. Casa Malaparte in Capri and Adalberto Libera", Lotus International, Vol. 60, 1989, p.9.
2. John Hejduk, "Cable from Milan", Domus, Vol. 605(April 1980), pp. 8~13.
3. Danielle Willems, "Cinematic Catalysts: Contempt + Casa Malaparte," The Funambulist, June 28th, 2011, http://thefunambulist.net/2011/06/28/essays-cinematic-catalysts-contempt-casa-malaparte-by-danielle-willems/.

永恒之屋
Alberto Campo Baeza

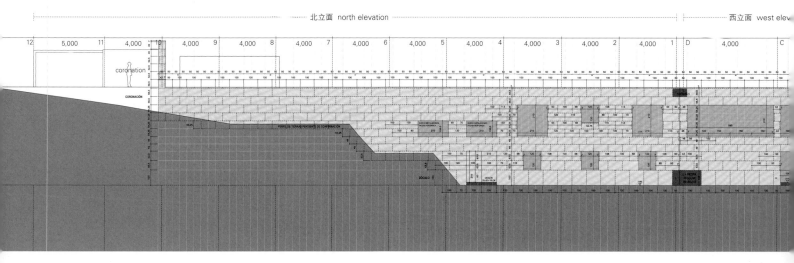

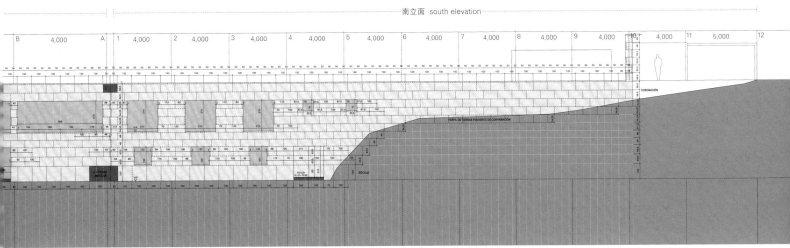

南立面 south elevation

我们在加的斯人间天堂般的神奇土地上修建了一座面朝无边大海的无限平面式居所。这是我们设计的最有特色的住宅。在大西洋水域边缘、新旧陆地相交的地方，一座石头平台横空跃起。所有来自地中海的船只都会经过这里，向着大西洋驶去。

我们在这儿建造的房子犹如伸向大海的码头，好似由顶部水平平面加冕的指挥台。在光滑裸露的巨大水平平面上，我们望向远方水天相接的地平线，欣赏夕阳西下的美景。这座如无边沙地般的罗马洞石之屋面对着广阔无垠的大海，不多一分，也不差毫厘。

为了实现这个升起的平台，我们建了一个巨大的盒子，宽20m，进深36m，是住宅的主要生活空间。平台下方前12m的区域内是两层的建筑结构，与平台一起构成了整个居住空间。

罗马人几个世纪前住在这里。博隆尼亚是罗马渔业工厂的废墟，人们曾经在这里生产咸鱼酱油，并给他们的神灵建寺庙，离我们建房子的地方只有咫尺之遥。为了纪念他们，我们用罗马洞石修建了这座房子，就像石头卫城。

为了扩大平台面积以更好地凸显平台元素，我们尽量将罗马洞石筑成的平台向后靠向入口处的墙，直逼街道。跨过高高的隔离墙，便会看到平台表面上住宅的入口：阶梯式的"壕沟"。

一位希腊诗人评价说这里是真正的圣地，是集会之所，是神话中人类和上帝朝会的地方。

裸露的石头平台上环绕着三面墙，可以抵御当地盛行的强风。有时，这些强劲的大风好似从风神的口袋中逃离出来，又像是推动尤利西斯的船踏上回归旅程的大风。

《基督早于人类出现》是伦布兰特1655年创作的一幅可爱的蚀刻版画。这幅画一直使我着迷。在这幅画中，伦布兰特勾画了一条笔直的水平线，很直很平。这条直线是强大的高台的边界，风景就在此平台上上演。正如密斯通常所做的那样，将平面变成了一条线。我确定伦布兰特和密斯一定会喜欢我们的平台房子，全是平台，只有平台。阿达尔贝托·利贝拉在设计马拉帕特别墅的时候也运用了同样的手法，他也一定会喜欢我们的平台房子。我们也很喜欢。当我们从沙滩看向我们的房子时，我们会想起所有这些人。

我们希望这个房子能够让时间静止，更希望这个房子永驻人们心间。这是一座永恒之屋。

House of the Infinite

On a marvelous place like a piece of earthly paradise, at Cádiz, we have built an infinite plane facing the infinite sea, the most radical house we have ever made. At the very edge of the waters of the Atlantic Ocean, where the sea unites the new and the old continent, emerges a stone platform. At the place where all the ships from the Mediterranean used to pass and still pass by as they head off into the Atlantic.

There we have erected a house as if it were a jetty facing out to sea. A house that is a podium crowned by an upper horizontal plane. On this resoundingly horizontal plane, bare and denuded, we face out to the distant horizon traced by the sea where the sun goes down. A horizontal plane on high built in stone, Roman travertine, as if it were sand, an infinite plane facing the infinite sea. Nothing more and nothing less.

To materialize this elevated horizontal plane, which is the main living room of the house, we built a large box with 20 meters of frontage and 36 meters deep. And under those first 12 meters we excavated two floors in the solid rock to develop the whole living space.

The Romans were there a handful of centuries ago. Bolonia, the ruins of the Roman fishing factories where they produced Garum and built temples to their gods, is just a stone's throw away. In their honor we have built our house, like an acropolis in stone, in Roman travertine.

To give even greater force to the platform we incorporated all the terrain as far back as the entrance wall separating us from the street, also done in Roman travertine. Once inside the wall,

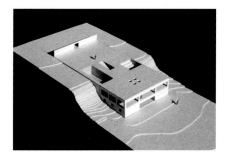

1 起居室	1. living room
2 门廊1	2. porch1
3 卧室	3. bedroom
4 浴室	4. bathroom
5 衣橱	5. closet
6 走廊	6. corridor
7 机房	7. machinery room
8 储藏室	8. storage room
9 桑拿室	9. sauna
10 土耳其浴浴室	10. hamman
11 通道	11. access
12 游泳池1	12. swimming pool1
13 门廊2	13. porch2
14 起居室/餐厅	14. living/dining room
15 主卧	15. main bedroom
16 更衣室	16. dress room
17 卫生间	17. W.C.
18 游泳池2	18. swimming pool2
19 厨房	19. kitchen
20 冷藏室	20. refrigerator room
21 洗衣房	21. laundry
22 家庭影院	22. home theater
23 电梯	23. elevator
24 停车场	24. parking
25 阁楼起居空间	25. pavilion living room
26 阁楼厨房	26. pavilion kitchen

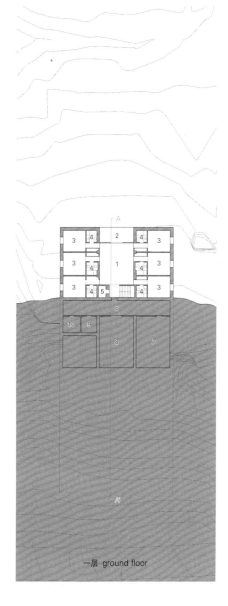

一层 ground floor

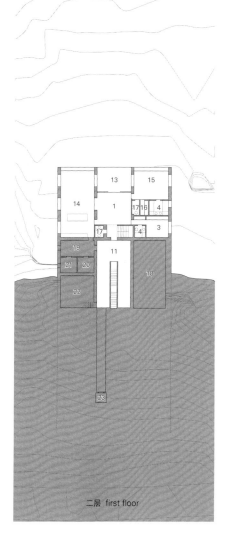

二层 first floor

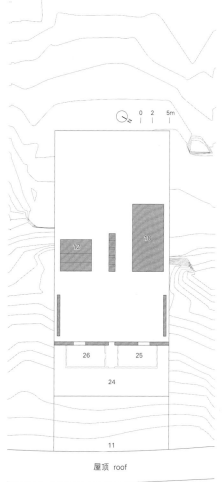

屋顶 roof

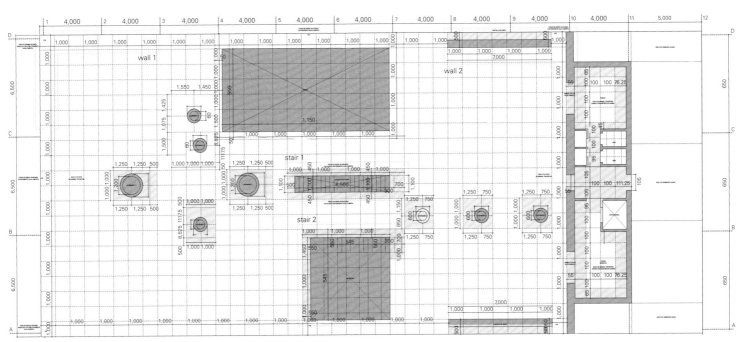

屋顶平台详图 roof platform detail

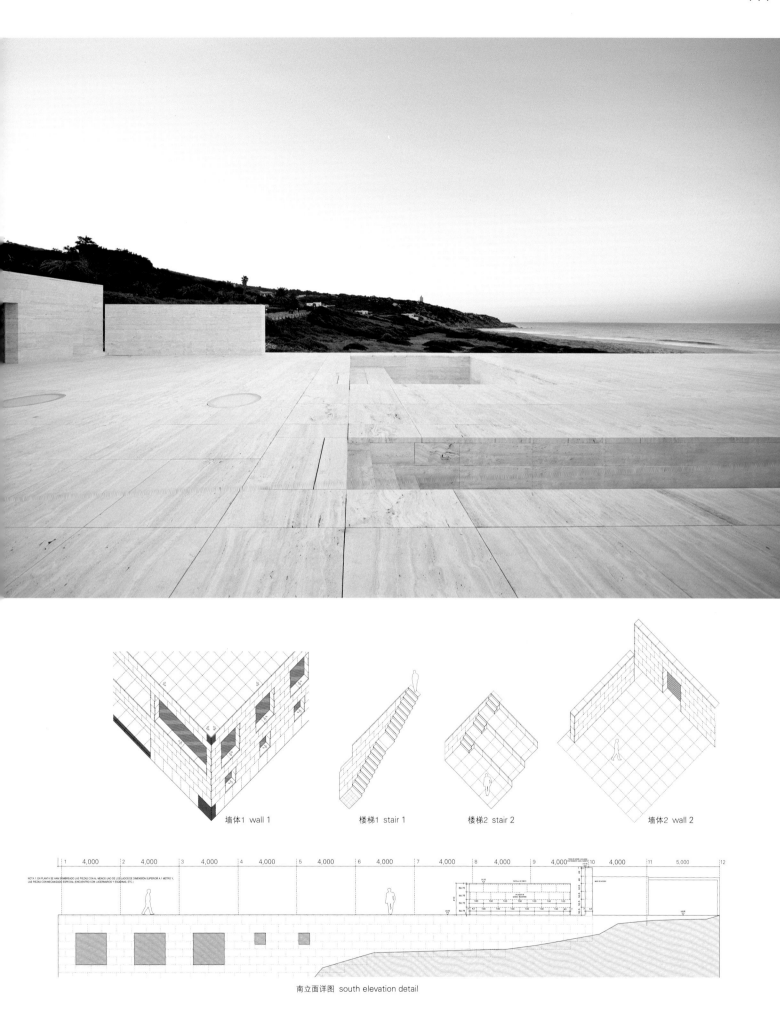

墙体1 wall 1　　楼梯1 stair 1　　楼梯2 stair 2　　墙体2 wall 2

南立面详图 south elevation detail

the entrance to the house will be via a "trench" in the form of stairs dug into the upper surface of the platform.

A Greek poet said that this is a true Temenos, a meeting-place, where according to mythology, humans and gods come together.

On the denuded stone platform, three walls surround us and protect us from the prevailing strong winds. Sometimes it is as if someone had opened the bag containing the winds of Aeolus. The same winds that drove on the vessel in which Ulysses made his journey home.

There is a lovely etching by Rembrandt from 1655, "Christ Presented before the People", that has always fascinated me. In it, Rembrandt sketches a straight horizontal line. Perfectly straight and perfectly horizontal. It is the border of the powerful dais, the podium upon which the scene takes place. There, as Mies did so often, he has made the plane into a line. I am certain that Rembrandt and Mies would like our podium house, all podium, only podium. As would Adalberto Libera, who did the same thing when he built his Malaparte House in Capri. And we like it too. And when we look at our house from the beach, we will be reminded of all of them.

We wanted this house to be capable not only of making time stand still, but to remain in the minds and hearts of humankind. The house of the Infinite. Alberto Campo Baeza

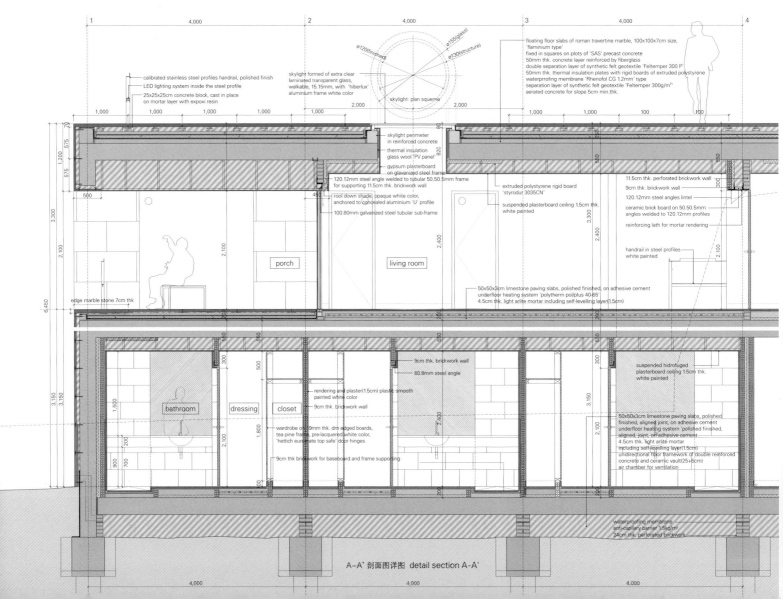

A-A'剖面图详图 detail section A-A'

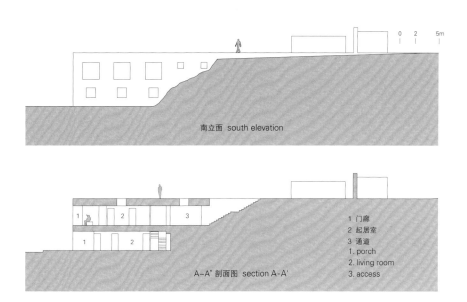

南立面 south elevation

1 门廊
2 起居室
3 通道
1. porch
2. living room
3. access

A-A' 剖面图 section A-A'

项目名称：House of the Infinite / 地点：Cádiz, Spain
建筑师：Alberto Campo Baeza
合作方：Tomás Carranza, Javier Montero, Alejandro Cervilla García, Ignacio Aguirre López, Gaja Bieniasz, Agustín Gor, Sara Oneto / 结构工程师：Andrés Rubio Morán
施工技术员：Manuel Cebada Orrequia / 质量控制：Laboratorios Cogesur
承包商：Chiclana
用地面积：3,500m² / 建筑面积：900m² / 总建筑面积：400m²
设计时间：2012 / 竣工时间：2014
摄影师：©Javier Callejas Sevilla (courtesy of the architect)

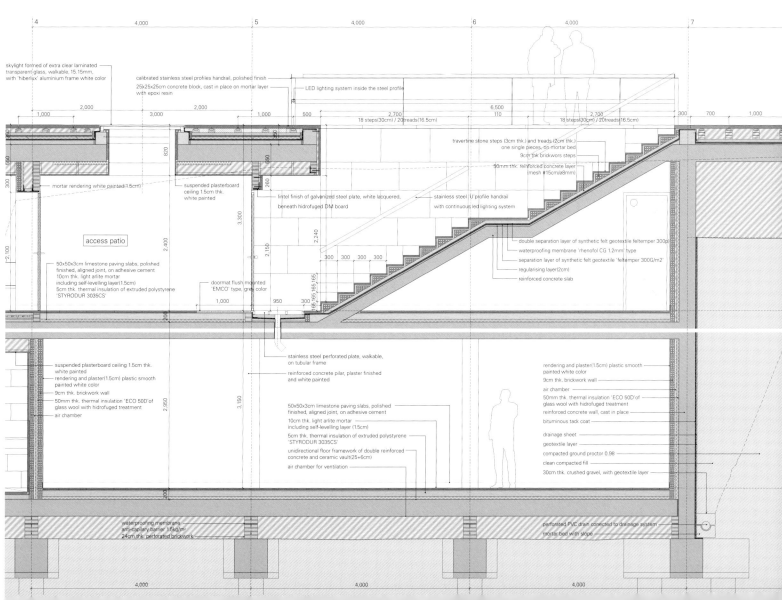

蒙桑双宅
João Paulo Loureiro

住宅设计–家居生态 Dwell How – Ecologies of Domesticity

/

在葡萄牙与西班牙交界处的蒙桑村附近，林木葱郁的米尼奥河岸边，是蒙桑双宅的所在。这个项目是为一对父子在同一地块建两栋住宅。儿子是当地有名的建筑商，准备接受我为他的新家提供的设计建议。因此，我决定将两座房子建在一个单体内。这好像很合逻辑。简而言之，两套住宅位于一块84m×16m的预应力混凝土板下，由既传统又新型的花岗岩墙支撑。整个设计充满现代气息（房子使用玻璃、漆铁和松木建造）。

这一现代而有力的建筑造型寻求建筑和周围自然景观的平衡与和谐。在长长的混凝土板的中心位置是一个圆形开洞，是一个椭圆。这里将来可以种一棵树、放置一件雕塑，或者仅仅是一处极具魅力的留白空间。对于在设计中将现代建筑技术与本地区传统的建筑技艺完美结合的不懈追求，实现了大跨度的水平状态和预应力混凝土的巧妙使用之间的对立统一，将钢筋混凝土发挥到极限。巨大厚重的花岗岩石块排列组合在一起，形成高墙，成为景观的一部分。

建筑形式和构造感觉都非常协调，与建筑周围环境融为一体，使建筑成了景观的组成部分。可以预见到所使用的材料作为项目的活跃因素在未来的反作用。建筑部分体量比如车库和室内泳池埋入地下，只在坡上露出细长的钢筋和预应力混凝土体量，似乎是破坏了所采用的伪装。在楼上，全部由玻璃镶嵌的铁框结构足以使自己消失在风景中，成为景观的一部分。建筑师所设计的一切，从铁制的窗框和门框，到乌木及松木打造的家具，虽然在这里构造主体是钢筋混凝土的，还是让人想起当地花岗岩教堂的祭坛装饰。为了保证建筑的能量平衡，采用了地热能，将整个管道系统都埋设在周围的大片土地中。

Two Houses in Monção

The site of the two houses is on the densely forested bank of the River Minho, the natural frontier between Portugal and Spain, near the town of Monção. The project was to design two houses on the same ground for father and son. The son is a well-known builder in the area. He was prepared for the proposal I was presenting for his new house. So, I have decided to join the two houses in only one body. It seemed logical. Briefly, the houses were made of a prestressed concrete slab (84m x 16m), a modern gesture to accommodate two houses (glass, painted iron and pine), supported on new albeit traditional granite walls.

It was kind of strong gesture seeking equilibrium with the landscape. At the center, a circle, an ellipse, a place for a tree to grow or for a sculpture or merely the space that remained and that conquers the site. The constant search to reconcile modern building technologies with the more traditional construction techniques of this region allows a dialectic between the horizontality of the large spans with the ingenious use of prestressed concrete, taking the reinforced concrete to its lim-

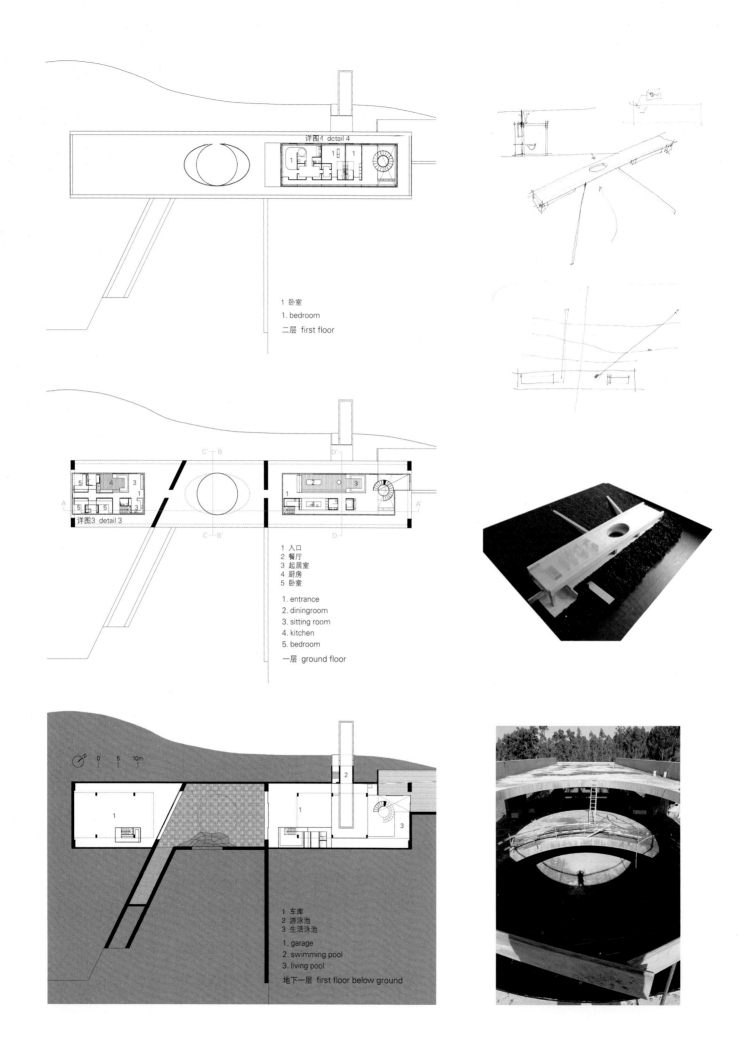

its, and the roughness of the large granite blocks which, juxtaposed, design large walls in the landscape.

The form and the tectonic feeling of the building were reconciled to establish a connection with the site, transforming it into an integral part of the landscape, foreseeing the future reaction of the materials as active dimension of the project. The building has a base buried into the earth, a garage and an indoor swimming pool, and is projected in a slender reinforced and prestressed concrete volume, over the slope, subverting the adopted camouflage. On the upper floor, the panes in a fully glazed iron structure are sufficient to dilute its mass in the landscape. Everything designed by the architect, from the iron window and door frames to the pine and ebony furniture, recalling the fit of the retables in the granite churches, although here set into reinforced concrete. Geothermal energy was adopted to ensure the energy balance of the building, which was achieved by burying all the piping systems in the large expanses of the surrounding land.

João Paulo Loureiro

项目名称：Two Houses in Monção
地点：Monção, Portugal
建筑师：João Paulo Loureiro
项目团队：Susana David Oliveira, Ana Luísa Cunha, Inês Caetano, Eva Vieira
结构工程师：Torção
后张法混凝土顾问：Fercanorte
用地面积：9,500m²
建筑面积：2,000m²
总建筑面积：3,900m²
设计时间：2008—2013
施工时间：2009—2013
摄影师：©José Campos (courtesy of the architect)

东立面 east elevation

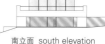
南立面 south elevation

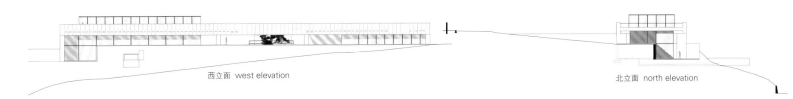

西立面 west elevation　　　　　北立面 north elevation

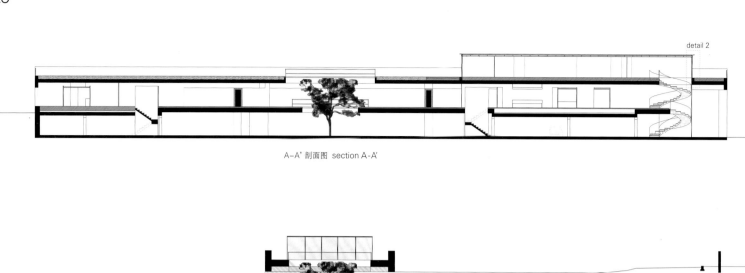

A-A' 剖面图 section A-A'

B-B' 剖面图 section B-B'

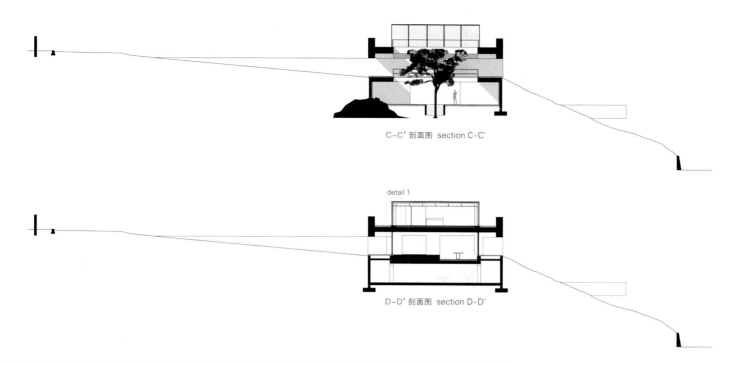

C-C' 剖面图 section C-C'

detail 1

D-D' 剖面图 section D-D'

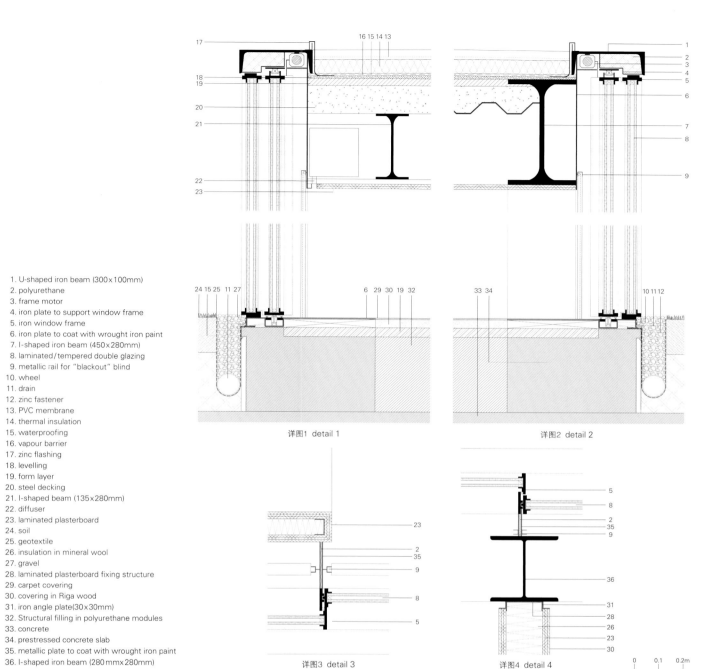

1. U-shaped iron beam (300×100mm)
2. polyurethane
3. frame motor
4. iron plate to support window frame
5. iron window frame
6. iron plate to coat with wrought iron paint
7. I-shaped iron beam (450×280mm)
8. laminated/tempered double glazing
9. metallic rail for "blackout" blind
10. wheel
11. drain
12. zinc fastener
13. PVC membrane
14. thermal insulation
15. waterproofing
16. vapour barrier
17. zinc flashing
18. levelling
19. form layer
20. steel decking
21. I-shaped beam (135×280mm)
22. diffuser
23. laminated plasterboard
24. soil
25. geotextile
26. insulation in mineral wool
27. gravel
28. laminated plasterboard fixing structure
29. carpet covering
30. covering in Riga wood
31. iron angle plate(30×30mm)
32. Structural filling in polyurethane modules
33. concrete
34. prestressed concrete slab
35. metallic plate to coat with wrought iron paint
36. I-shaped iron beam (280mm×280mm)

详图1 detail 1

详图2 detail 2

详图3 detail 3

详图4 detail 4

卡特彼勒住宅
Sebastián Irarrázaval

这座预制住宅是为了一个艺术收藏家及其家人而建的，位于圣地亚哥郊外一个新建的住宅区。为了减少施工时间和成本，使用了二手集装箱：5个12.2m的标准集装箱，6个6.1m的标准集装箱以及一个顶部开口的12.2m集装箱用于建游泳池。

先不谈实现设计标书中的要求，建造该住宅的两个主要目的是：

首先，将房屋与所在城市的这部分地域加以整合。在这个地方，安第斯山脉表现出了强烈的视觉和地质构造上的冲击力。因此，安第斯山脉作为一个明显的背景最好不要被突出强调，另外，对于倾斜地面的设计也需要协商进行。

其次，为了避免采用机械制冷方式，要使室外空气在整栋房屋和各个组成部分中顺畅并轻松地流通。

对于前者，为了达成与场地整合的设计，所采取的策略是，在摆放这些集装箱建筑体量时，仿佛是依靠在斜坡上，也就是说，让这些建筑体量与斜坡融为一体。房屋的入口设计和儿童卧室的设计都是房屋与自然地势适应调整的结果。每间儿童卧室都有倾斜的室内空间，也是安装天窗和放置床的地方。

至于后者，为提升空气在房屋中的流通，策略是以带状形式组织房屋结构，各个空间之间的缝隙空间既可以用作居住者进出的通道，也可以带来安第斯山脉的凉爽空气。同时，缝隙空间增加了房屋的周长，让房屋大部分时间都能够拥有来自至少两个相反方向的光和空气进入这些空间。为了达到上述目的，窗和门都沿垂直于条带的轴线排成一行，这样的话，空气流通更容易，也创造了视觉上的整合统一。

施工阶段包括：首先，建造挡土墙，以创建一个水平面来设置房屋的公共部分；第二，安装集装箱，并安装在顶层作为隐私区域；第三，将集装箱外表包裹上统一的材料，既使房屋的各个部分看起来整齐划一，又使外立面通风换气，有助于调控室内空间的温度。

建筑材料的色彩选择既考虑降低材料成本，也考虑减少维护成本。在选择材料的时候，分析它们的耐久性也同样重要，将时间因素作为材料的附加值来考虑。

建筑的基本构造元素如窗、门和天窗的设计都合情合理，在整个房屋里都是一样的，这样做不仅可以减少成本，也为了营造出整个建筑构造的和谐宁静。

Caterpillar House

This prefabricated house for an art collector and his family was built in the outskirts of Santiago in a new suburban residential area. In order to reduce construction time and costs, second hand shipping containers were used as follow: five 40" standard containers, six 20" standard containers and one 40" open top container for the swimming pool.
Putting aside the fulfillment of the brief; the main purposes of the house were two:

The first one was to integrate it to the territory of this part of the city where the presence of the Andes Mountain is extremely strong both visually and tectonically. Therefore, the presence of the Andes was considered as an obvious background worst to be highlighted and also as a sloped ground needed to be negotiated.
The second one was to allow the external air to run smoothly and easily through all the house and its different parts in order to avoid mechanical cooling.

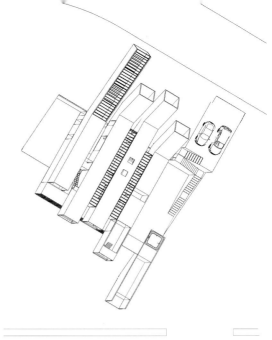

二层 second floor

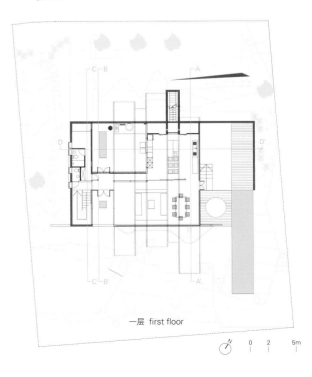

一层 first floor

0　2　5m

With regard to the former, the strategy to achieve the proposed integration to the territory, consisted in placing the volumes of the house as if they were resting on the slope and allowing the volumes, let's say; to be blended by the slope. As a result of this tide adjustment of the house to the natural ground, there generated the entrance to the house and the children sleeping rooms having each one an inclined inner space which is both the skylight and the bed place in the room.

In relation to the latter; the strategy for improving air move-

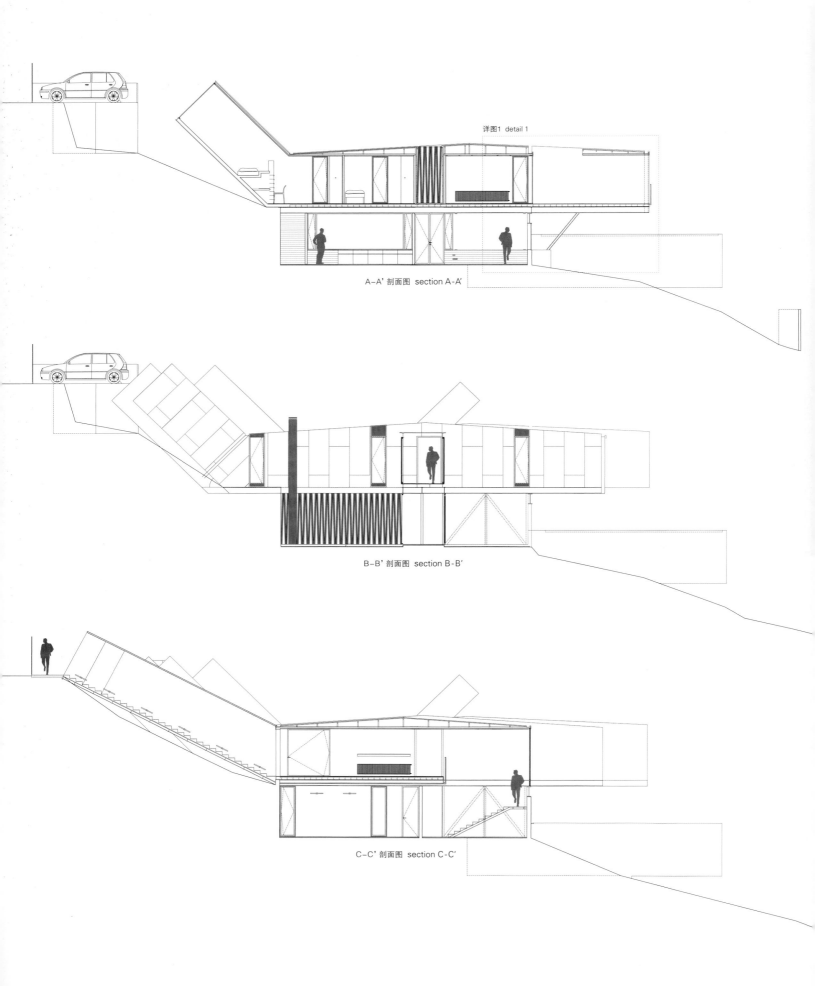

详图2 detail 2 D-D'剖面图 section D-D'

ment through the house consisted in organizing the program along stripes and keeping interstitial spaces among them for the circulation of both; the inhabitants and the cool air that comes from the mountains. At the same time, the proposed interstitial spaces increase the house perimeter which allows most of the times to have light and air entering the spaces from at least two opposite sides. As a consequence of this will, windows and doors are aligned along axes that cut the stripes, therefore easing air movement and also creating visual integration.

Construction phases consisted of: firstly; placing the retaining walls to create a horizontal plane to place the public areas of the house. Secondly; in mounting the containers and ensemble them on top to place the private areas and thirdly; to wrap the containers with a Unitarian material which at the time that integrates the parts also creates a ventilated facade that temperate the interior spaces.

The material palette was chosen having in mind not only low cost materials but low maintenance as well. In the election of materials it was also important to analyze their capacity to age well and to incorporate time passing as something that adds value to material.

Architectonic elements such as windows, doors, and skylights are rationalized and repeated all over the house not just to reduce cost but also in order to create an integrated architectonic peace.

项目名称：Caterpillar House
地点：Los Trapenses , Lo Barnechea, Santiago de Chile
建筑师：Sebastián Irarrázaval Delpiano
参与设计建筑师：Erick Caro. / 结构工程师：Pedro Bartolomé
施工：Sebastián Irarrázaval Arquitectos / 场地监理：Ricardo Carril
材料：5×40" standard shiping containers, 6×20" standard shiping containers, 1×40" open top shiping container, steel plates, concrete retaining walls, gypsum board, tymber wood
客户：Ricardo Bezanilla / 用地面积：900m² / 总建筑面积：350m²
竣工时间：2012
摄影师：©Sergio Pirrone

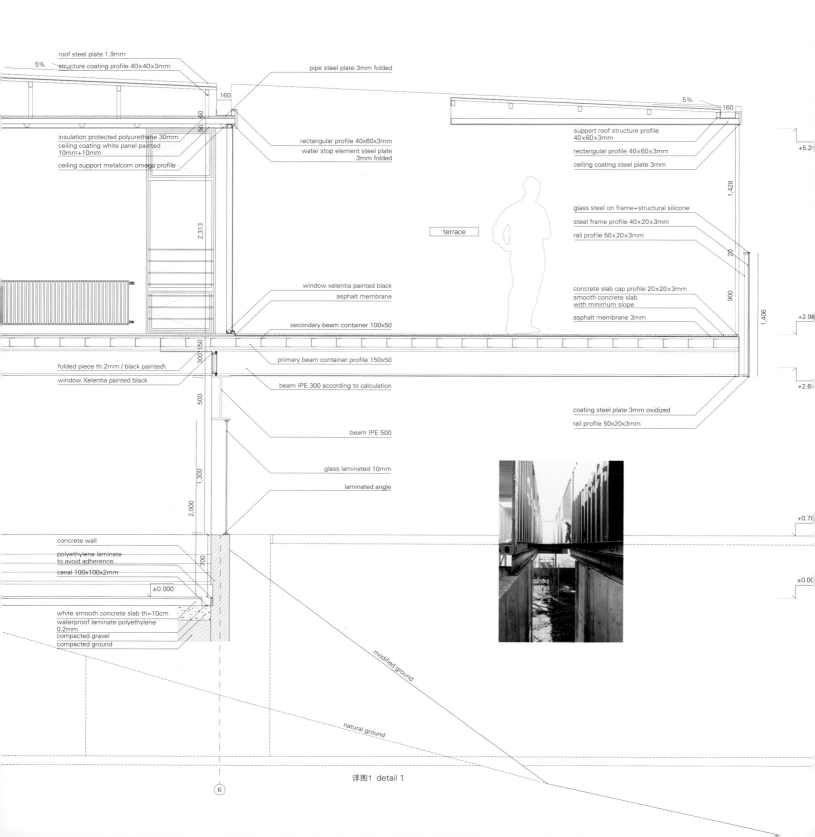

详图1 detail 1

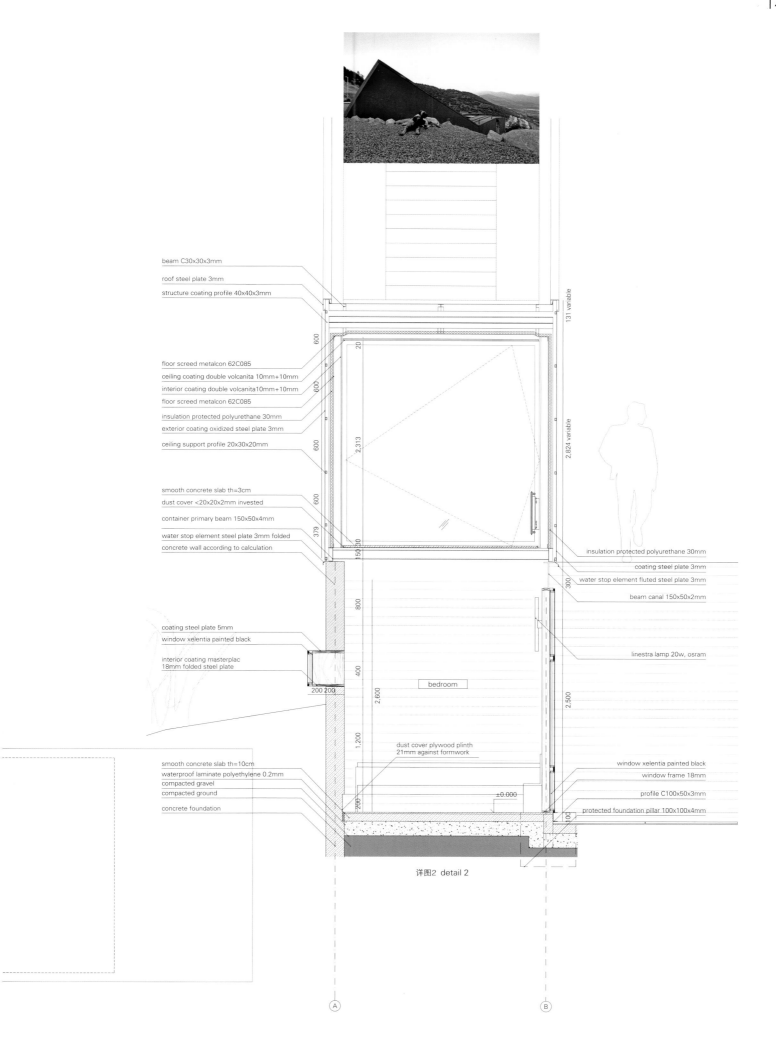

详图2 detail 2

八岳山住宅
Kidosaki Architects Studio

这座住宅坐落在八岳山山脚下一块倾斜的山脊上，从这里可以欣赏到鲜为人知的壮观景色。为了寻找最美的如画般的风景，客户在立科町住了下来，花了多年时间去寻找理想的地点来建造他的房子。不可避免的是，这个项目的主要目的是满足客户的期望，把这些令人惊喜的美景融入到设计中。

当我考察项目地点时，我的第一印象是，必须将这里未经开发的广袤的大自然尽可能地包含到建筑内部。建筑师决定以能够最大化欣赏水平延伸的美景为前提来安排住宅的布局。为了实现这样的设计，设计师采用了巨型结构柱，使住宅的一半水平延展在空中。巨大的悬挑楼板通过两根斜撑钢柱对角支撑，每根钢柱的直径都是300mm。这样一来，住宅就像是飘浮在周围壮丽的自然风景中一样。悬挑结构也让高原山区气流带动的微风穿过室内，住宅与大自然共存。

当你受邀进入住宅时，穿过围挡的走廊，就会进入一个扣人心弦、令人激动不已的空间，壮观且令人印象深刻的景色会展现在你的眼前。在客厅、餐厅和厨房区域的三个方向，壮观的全景一望无际。这是你在其他任何地方都无法找到的美景，只有在这个空间才能欣赏到。这里的风景都是属于你自己的。这个空间会给人一种奢侈的体验，只有那些被邀请的人才能真正享受到。从其他房间能看到山脉的不同景色，每个房间的风景各不相同。

高高的天花板、宽阔的木甲板和突出的屋檐都使整个空间沉浸在这美不胜收的全景中。这种感觉是如此强烈，就好像生活在云朵上一样。

被抬高的各个建筑组件都经过精心的细节设计，精致的结构赋予空间一种张力感和统一感，恰当的建筑材料运用在建筑作为主导因素和与周围自然环境和谐一致之间实现了适当的平衡。

这座住宅按照建筑师的设计愿景建造，其特色和谦恭的姿态表达了对周围景色虔诚的敬重。

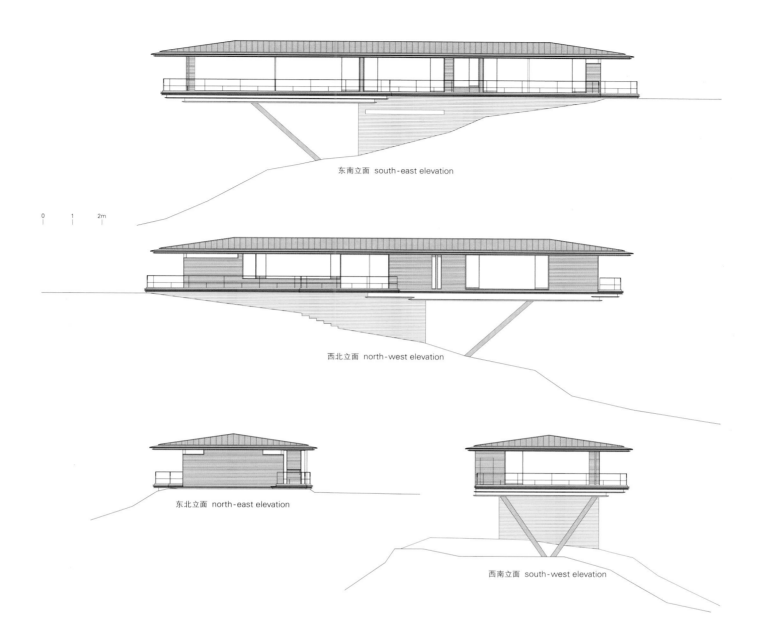

东南立面 south-east elevation

西北立面 north-west elevation

东北立面 north-east elevation

西南立面 south-west elevation

House in Yatsugatake

Located on a sloping mountain ridge at the foot of the Yatsugatake Mountains, this house was designed on a piece of land that offers spectacular views that are rarely known. Seeking for the best in picturesque scenery, the client took up residence in Tateshina, and spent many years searching for the ideal site for building his house. Inevitably, the main aim of this project is to meet the client's expectations to incorporate these stunning views into the design.

When I visited the site, my first impression was that this untapped and expansive nature must be embraced into the interior to the greatest extent possible. The architect decided to arrange the house in such that this horizontal expanded scenery must be maximized. In order to realize this design, columns were introduced as mega structures, enabling half of the house to extend into the air. A large overhanging floor, 2 diagonal bracing steel cylinders, each 300mm in diameter is introduced. With this, the house floats in the midst of a glorious nature. And with this overhanging structure, the breeze of the mountain plateau flows through the interior, creating a coexistence with nature.

When you are invited into the entrance way, after passing through the restrained space of the hallway, and a dramatic space with magnificent and impressive scenery unfold. A

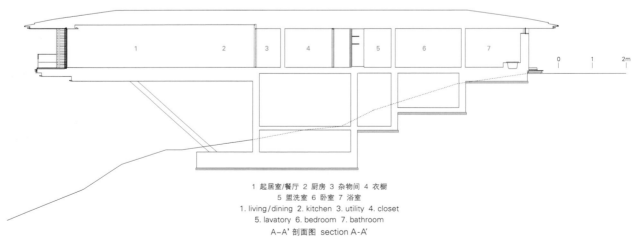

1 起居室/餐厅 2 厨房 3 杂物间 4 衣橱
5 盥洗室 6 卧室 7 浴室
1. living/dining 2. kitchen 3. utility 4. closet
5. lavatory 6. bedroom 7. bathroom
A-A' 剖面图 section A-A'

项目名称：House in Yatsugatake / 地点：Nagano, Japan / 建筑师：Kidosaki Architects Studio / 总建筑师：Hirotaka Kidosaki / 项目团队：Yuko Sano / 结构工程师：Takashi Manda, Mitsuru Kobayashi _ Toyohito Shibamura Structure Design / 总承包商：Niitsugumi / 结构系统：Reinforced concrete, steel-frame partly / 用地面积：2,044.68m² / 建筑面积：303.00m² / 总建筑面积：199.80m² / 施工时间：2011.9—2012.7 / 摄影师：©45g Photography- Junji Kojima(courtesy of the architect)

majestic panoramic view extends on all three sides of the living, dining, and kitchen area. This is something you can't find anywhere else but here in this space. The scenery is all to one's own. And only those who have been given a privilege to be invited can truly enjoy the extravagant experience of this space. Other rooms are planned to offer various different views of the mountains from each of the rooms.

The high ceilings, wide wood deck, and protruding eaves enable a space steeped in the overwhelming presence of the panoramic views of the area. The feeling is so intense that it is almost as if you are living on a cloud.

The various components have been elevated through careful attention to detailing. And the refinement of the structure gives a sense of tension and unity to the space. And its adequate materials achieving the proper balance between a dominance and a harmony with the surrounding natural environment.

The character and humility of this dwelling, constructed without compromising the vision of the architect, expresses a dignified reverence for the scenery surrounding it.

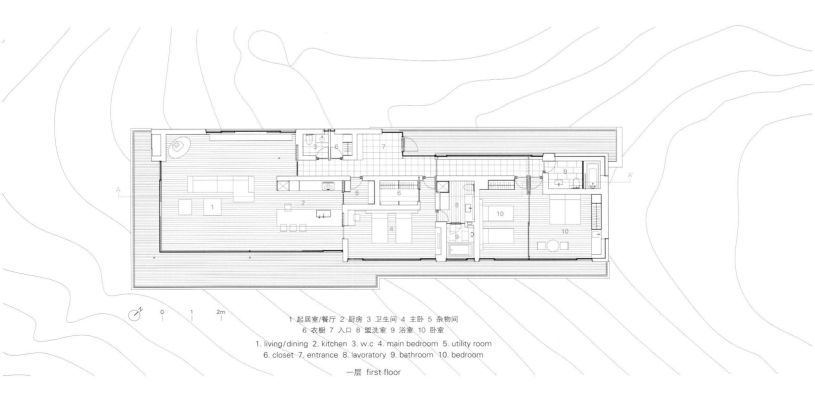

1 起居室/餐厅 2 厨房 3 卫生间 4 主卧 5 杂物间
6 衣橱 7 入口 8 盥洗室 9 浴室 10 卧室

1. living/dining 2. kitchen 3. w.c 4. main bedroom 5. utility room
6. closet 7. entrance 8. lavoratory 9. bathroom 10. bedroom

一层 first floor

MO住宅
Gonzalo Mardones Viviani

MO住宅位于一个陡峭的斜坡上，面朝智利萨帕利亚尔海岸。房屋建于公路地平线以下，丝毫无意破坏从公路地平线方向而来的美妙视野。所以，需要通过楼梯和坡道才能向下进入山中再次探寻空间，去发现MO住宅、大海、海岸和冲到海滩上的浪花。

住宅的入口位于中间一层，这里设计有公共区域，包括客厅、餐厅和厨房。卧室建在了一层，从这里可以直接进入花园。主卧室被设计在了上面一层，位于悬挑的体量之上，拥有良好的视野，突出了房屋设计的明显意图：捕捉地平线的美景。

悬浮于空中的建筑体量在设计中至关重要，并足以证明房子本身就是一个巨大的X轴。房子的每一层都有凉台，巨大的屋檐和横梁可以为凉台提供保护，遮挡西晒与烈风。

加入二氧化钛的钢筋混凝土是这栋住宅所使用的唯一建筑材料。钢筋混凝土的使用使得房屋地基埋在地下，房体建造在斜坡上，而且在智利这样的地震带上的国家也能拥有轻盈的结构。在混凝土中加入二氧化钛可以使混凝土颜色变浅，同时像树木一样能消除车辆产生的有毒气体。整个房子的内部都被涂成了白色，有意强化室内的自然明亮，大量地运用向地平线延伸出去的窗户、天窗、采光庭院和通风换气来塑造明亮的室内空间。门窗框等木工建材均采用雪松。

屋顶经过精心的设计，用白色石头装饰，成为房屋的第五立面，从外面的公路看过去就是房屋的主立面。外立面所使用的石头的白色就如同海边翻涌的白色浪花。另外，白色石头装饰的屋顶和石头下面的混凝土结构之间有一个空气层，能够有效避免建筑内部受到阳光直射，更利于通风换气，进而调节了住宅的室内温度。

MO House

MO House is located at an area of steep slope facing the sea coast of Chile in the commune of Zapallar. The house is buried with clearly no intention to interrupt the wonderful view to the horizon from the public road, so it is accessed by stairs and ramps going deep into the mountain to discover again, from inside the house, the sea, coast and breaker.

The access to the house is in the middle level, which contains the public areas of the house: living room, dining room and

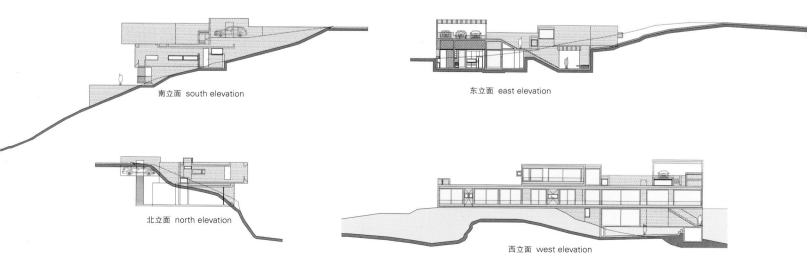

南立面 south elevation

东立面 east elevation

北立面 north elevation

西立面 west elevation

1 户外烧烤场地	1. barbecue
2 游戏室	2. game room
3 储藏室	3. storage room
4 浴室	4. bathroom
5 游泳池	5. pool
6 入口门厅	6. entrance hall
7 卧室	7. bedroom
8 起居室	8. living room
9 客房浴室	9. guest bathroom
10 设备室	10. service room
11 厨房	11. kitchen
12 餐厅	12. dining room
13 露台	13. terrace
14 洗衣房	14. laundry
15 主卧	15. main bedroom
16 车库	16. parking

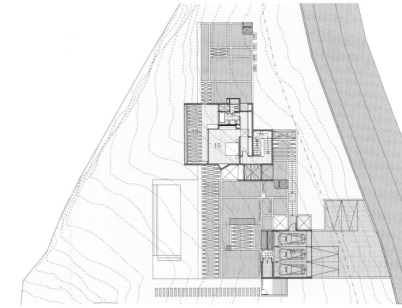

二层 first floor

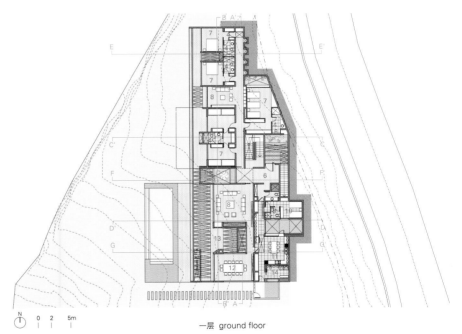

一层 ground floor

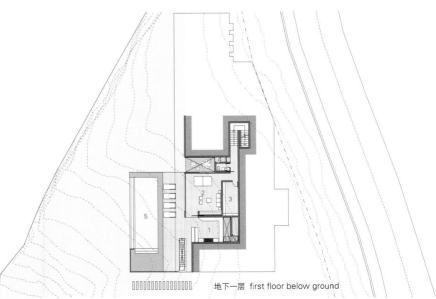

地下一层 first floor below ground

项目名称：MO House
地点：Zapallar, Chile
建筑师：Gonzalo Mardones Viviani
工程师：Ruiz y Saavedra
照明设计师：Paulina Sir
用地面积：2,380m²
建筑面积：489.98m²
设计时间：2011—2012
施工时间：2013—2014
竣工时间：2014
摄影师：©Nico Saieh (courtesy of the architect)

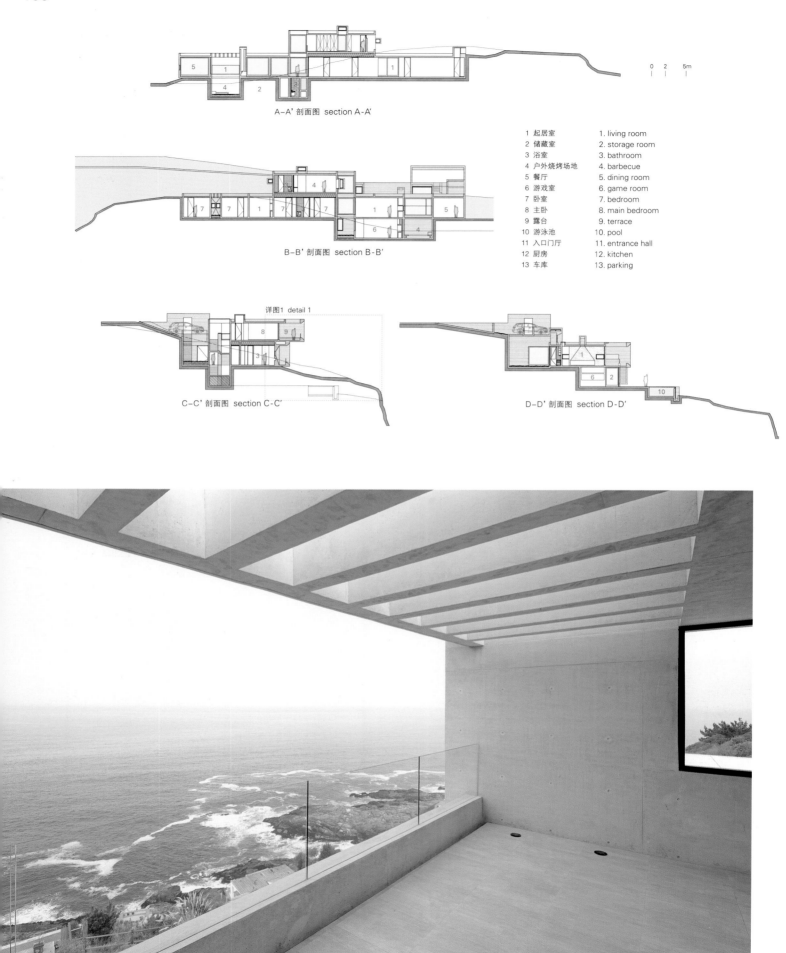

A-A' 剖面图 section A-A'

B-B' 剖面图 section B-B'

详图1 detail 1

C-C' 剖面图 section C-C'

D-D' 剖面图 section D-D'

1 起居室	1. living room
2 储藏室	2. storage room
3 浴室	3. bathroom
4 户外烧烤场地	4. barbecue
5 餐厅	5. dining room
6 游戏室	6. game room
7 卧室	7. bedroom
8 主卧	8. main bedroom
9 露台	9. terrace
10 游泳池	10. pool
11 入口门厅	11. entrance hall
12 厨房	12. kitchen
13 车库	13. parking

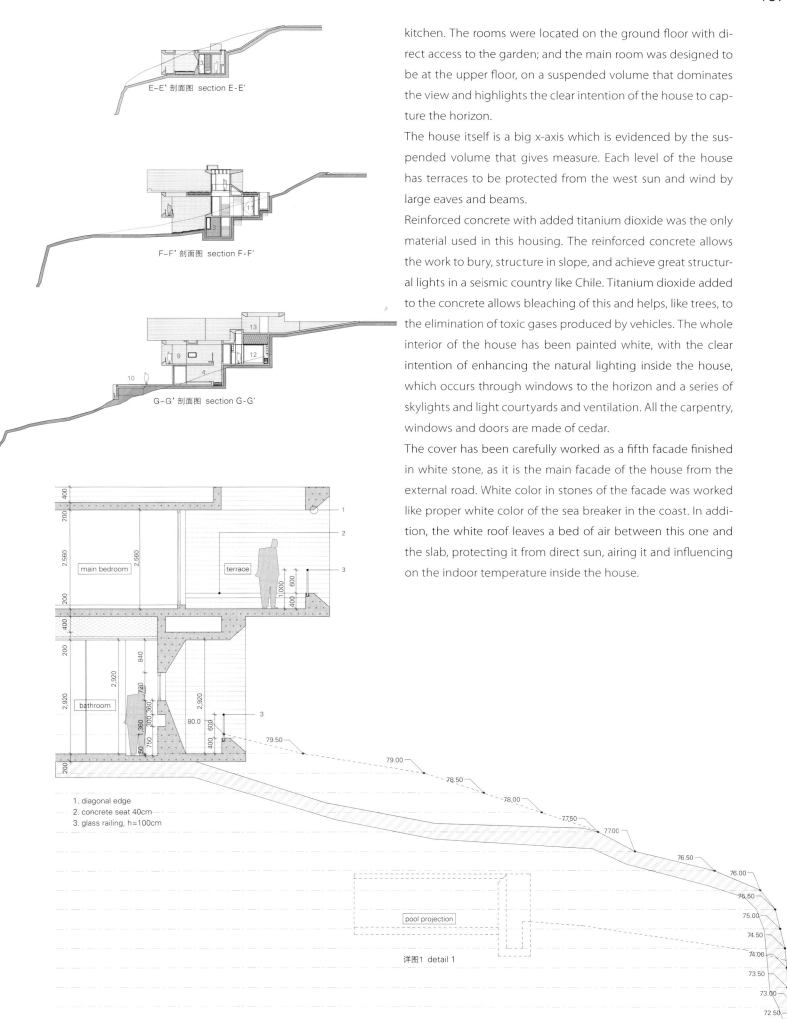

1. diagonal edge
2. concrete seat 40cm
3. glass railing, h=100cm

详图1 detail 1

kitchen. The rooms were located on the ground floor with direct access to the garden; and the main room was designed to be at the upper floor, on a suspended volume that dominates the view and highlights the clear intention of the house to capture the horizon.

The house itself is a big x-axis which is evidenced by the suspended volume that gives measure. Each level of the house has terraces to be protected from the west sun and wind by large eaves and beams.

Reinforced concrete with added titanium dioxide was the only material used in this housing. The reinforced concrete allows the work to bury, structure in slope, and achieve great structural lights in a seismic country like Chile. Titanium dioxide added to the concrete allows bleaching of this and helps, like trees, to the elimination of toxic gases produced by vehicles. The whole interior of the house has been painted white, with the clear intention of enhancing the natural lighting inside the house, which occurs through windows to the horizon and a series of skylights and light courtyards and ventilation. All the carpentry, windows and doors are made of cedar.

The cover has been carefully worked as a fifth facade finished in white stone, as it is the main facade of the house from the external road. White color in stones of the facade was worked like proper white color of the sea breaker in the coast. In addition, the white roof leaves a bed of air between this one and the slab, protecting it from direct sun, airing it and influencing on the indoor temperature inside the house.

"秋之屋"
Fougeron Architecture

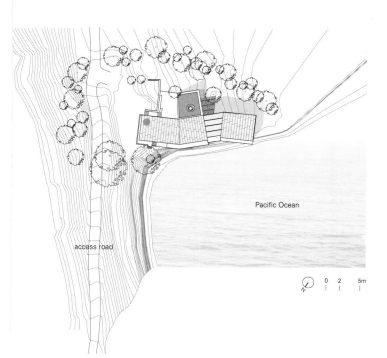

将房屋建在荒野是很激进的行为。本项目所面临的挑战是将房子建在太平洋沿岸最壮观的自然景色中,既要尊重这块土地,又要改变这块土地。

这座三室住宅位于大索尔壮观的南海岸线,扎根于加利福尼亚景观的天然美感与力量中。该建筑的设计策略是将建筑嵌入场地,创造一个和周围环境密不可分的结构。场地风景优美:断崖和房屋西侧与太平洋有76m的落差。因此,人们需要的是一个更加复杂的外形设计而不仅仅是一个巨大的风景窗。

细长的体量就像当地海滨森林里特有的香蕉蛞蝓一样,既遵从自然的地形轮廓和断崖绝壁的几何外形,又改变了它们。这样,复杂的房屋结构系统既利用自然的地形地貌,又不完全遵守自然的地形地貌,以完全融合地形。房屋伸出断崖3.7m,既保护悬崖微妙而脆弱的生态系统,又保证了结构的完整性和安全性。与广袤而波涛汹涌的大海和乱石嶙峋的悬崖形成对比,室内就是一处避风港。房屋设计也使得南面的室外空间免受强劲的西北风的侵袭。

房屋的主体是由两个矩形盒子组成，中间由全透明的玻璃图书室/小房间连为一体。主入口位于上层体量的顶部，起居空间从顶部最开放的公共空间到下面一层最私密的空间逐一呈现。起居室、厨房和餐厅为敞开式设计，为了区分不同的功能空间，楼层与屋顶平面都做了细微的改变。

较低一层的体量设有双悬臂式主卧室套房，如同海上凸起的海角，人们在落地窗前可以看到壮观的醉人美景。两个建筑体量的连接部分是玻璃图书室/小房间，它是房子的核心，其几何结构和全透明设计将房屋内外连为一体。

垂直于房屋且只有一层的用混凝土建的侧翼结构内设一间卧室和房屋设备间，屋顶进行了绿化。该结构的作用就是将房屋紧紧锁在大地上的巨石。

房屋有两个主要外立面。南外立面的墙和屋顶用铜皮包裹。铜覆层屋顶的悬臂结构保护窗户和前门免受风吹日晒。北外立面都是玻璃的，透明开阔的玻璃面使房子拥有无限视野。

Fall House

Placing form on wilderness is a radical act. The challenge was to design a house in one of the most spectacular natural settings on the Pacific Coast that would both respect and transform the land.

This three-bedroom home, on Big Sur's spectacular south coast, is anchored in the natural beauty and power of this California landscape. The design strategy embeds the building within the land, creating a structure inseparable from its context. The site offers dramatic views: a 250-foot drop to the Pacific Ocean both along the bluff and the western exposure. Yet it demands a form more complex than a giant picture window. The long, thin volume conforms and deforms to the natural

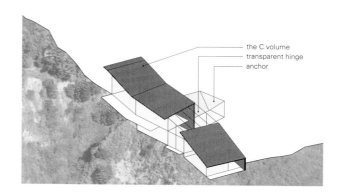

- the C volume
- transparent hinge
- anchor

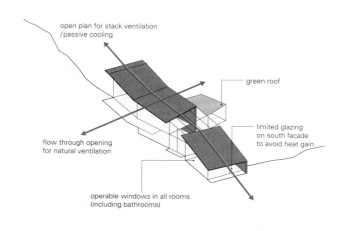

- open plan for stack ventilation / passive cooling
- green roof
- limited glazing on south facade to avoid heat gain
- operable windows in all rooms (including bathrooms)
- flow through opening for natural ventilation

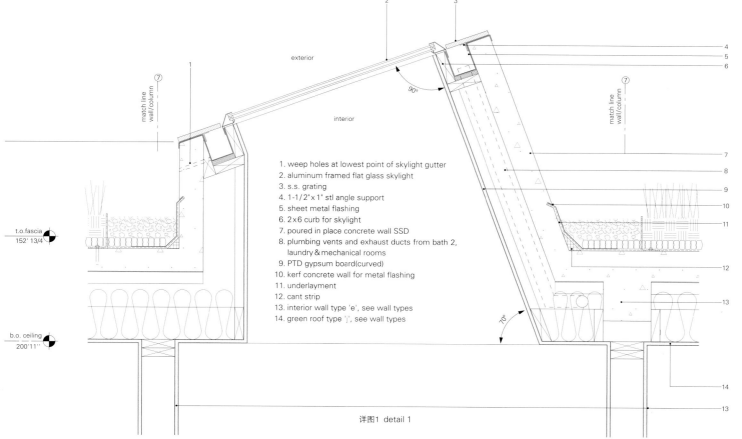

1. weep holes at lowest point of skylight gutter
2. aluminum framed flat glass skylight
3. s.s. grating
4. 1-1/2" x 1" stl angle support
5. sheet metal flashing
6. 2×6 curb for skylight
7. poured in place concrete wall SSD
8. plumbing vents and exhaust ducts from bath 2, laundry & mechanical rooms
9. PTD gypsum board(curved)
10. kerf concrete wall for metal flashing
11. underlayment
12. cant strip
13. interior wall type 'e', see wall types
14. green roof type 'j', see wall types

详图1 detail 1

项目名称：Fall House
地点：Big Sur, California, USA
建筑师：Fougeron Architecture
结构工程师：Endres Ware Architects Engineers
景观建筑师：Blasen Landscape Architects
土木＆土工技术工程师：Grice Engineering and Geology, Inc.
用地面积：510.97m² / 总建筑面积：353.03m² / 有效楼层面积：353.03m²
设计时间：2009 / 竣工时间：2013
摄影师：©Joe Fletcher Photography (courtesy of the architect)

二层 first floor

一层 ground floor

1 起居室	1. living room
2 厨房/餐厅	2. kitchen/dining
3 图书馆	3. library
4 主卧室	4. master bedroom
5 卧室	5. bedroom
6 储藏室	6. storage
7 露台	7. patio
8 平台	8. deck

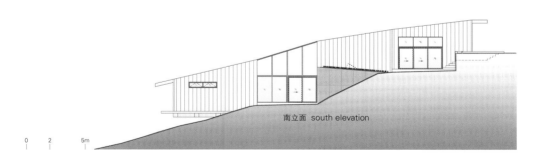

南立面 south elevation

0 2 5m

北立面 north elevation

contours of the land and the geometries of the bluff, much like the banana slug native to the region's seaside forests. In this way, the complex structural system applies and defies natural forms to accommodate the siting. The house is cantilevered 12 feet back from the bluff, both to protect the cliff's delicate ecosystem and to ensure the structure's integrity and safety. The interior is a shelter, a refuge in contrast with the roughness and immense scale of the ocean and cliff. The house also shields the southern outdoor spaces from the powerful winds that blow from the northwest.

The main body of the house is composed of two rectangular boxes connected by an all-glass library/den. The main entry is located at the top of the upper volume with the living spaces unfolding from the most public to the most private. The living room, kitchen and dining room are an open plan with subtle changes in levels and roof planes to differentiate the various functions.

The lower volume, a double-cantilevered master bedroom suite, acts as a promontory above the ocean, offering breathtaking views from its floor-to-ceiling windows. The link between these two volumes is the glass library/den; it is the hearth of the house, a room that unites the house inside and out both with its geometry and its transparency.

A one-story concrete wing perpendicular to the house includes a ground-floor bedroom, building services and a green roof; it is the boulder locking the house to the land.

The house has two main facades, the south one is clad in copper which wraps up the wall and over on the roof. Copper clad roof overhangs protect windows and the front door from the sun and the wind of the ocean. The facade to the north is made all glass; clear expanses of glass open the house to the view.

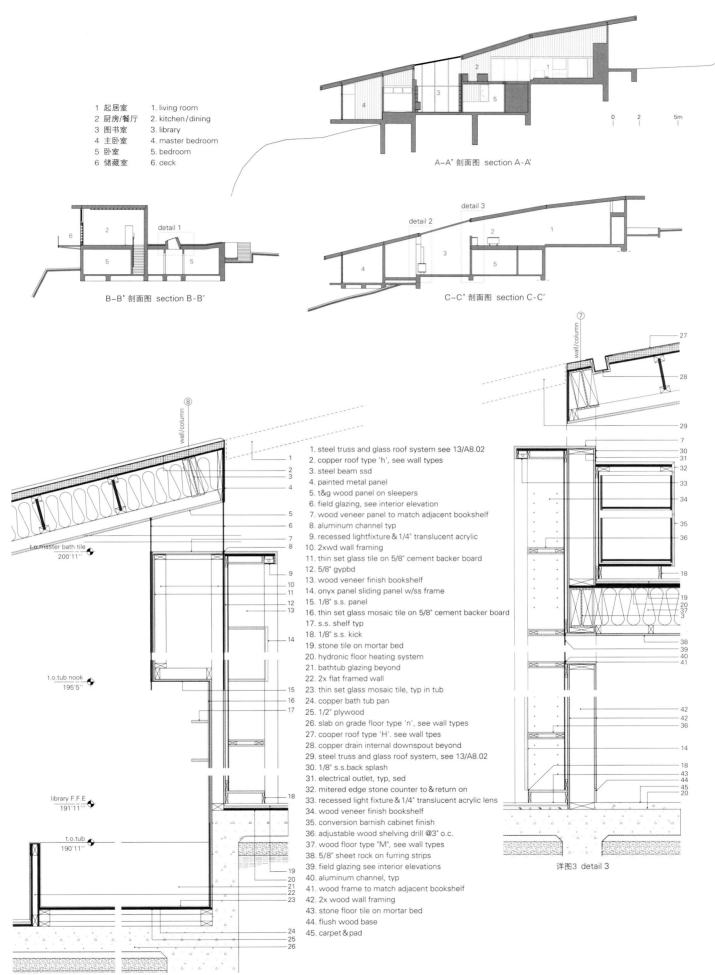

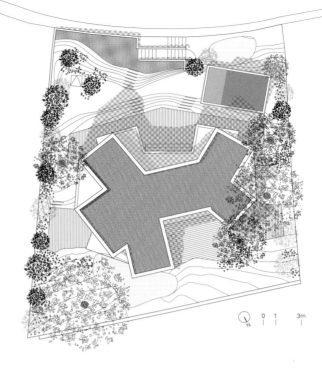

向日葵住宅
Cadaval & Solà-Morales

向日葵住宅坐落于一个条件受限的特殊地带。在地中海的海水和科斯塔布拉瓦的坚硬岩石组成的边界内，克莱乌斯角自然公园的自然荒野和埃尔波尔特德拉塞尔瓦的城市居住区之间，一个法国和西班牙边境的小渔村坐落于此，是比利牛斯山脉延伸入海的地方。这里无论是在海岸和水中，野生动植物资源都异常丰富。这个住宅想要充分利用这些壮观风景的特色，其几何结构设计使住宅形成不同的特定视野，同时扩大空间，让视野更加广阔开放。

梅尔和杰夫想要一个能够看到全景并且面向地中海的住宅。但他们从没有考虑到，场地超出其亲近大海的距离，会遭受到这个半岛最强烈的风暴之一（Tramuntana海风，在此可达180公里/小时）的侵袭，并且几乎得不到任何直接的太阳照射。所以该项目设计从两个方面开始考虑：加强与大海的联系，同时寻找将阳光引进屋内的方法。

从住宅的正面望出去，其视野令人印象深刻，从法国延伸到西班牙的Cabo de Creus自然公园，广袤浩瀚的大海就在眼前，无边无际。人们还可以欣赏到岩石以及由于风吹而不断改变的天空。该项目将全景变成不同条件下的局部景观；该住宅的不同用途空间先被细化然后连为一体，这样，每个部分（都是小规模）都面朝着先前枚举的不同景色。因此，该项目就是一个个小单元组成的空间，每个小单元都有不同的视野，人们就在从一个小单元向另一个小单元的过渡中获得了全景视野。也是这些增建的小单元空间组合在一起才形成一个主要的开放空间，即整个房屋的中心空间。

该住宅也是一个巨大的太阳能集热器，如同一个巨大的向日葵，一个可以把光和热引进屋内的机械系统。不同建筑模块围绕着后部的天井，天井将太阳光线引进客厅，让整个住宅温暖起来。该天井本身利用自身的结构可以免受强劲的Tramuntana风的影响，其朝向可以确保将最大限度的太阳光引进屋内。当Tramuntana风袭击这个地区时，这也是一处人们可以停留的户外空间。并且，人们从这个位于房屋后部、四周都是岩石和当地植被的天井方向仍然能通过两大玻璃窗欣赏到海景。

分割成小单元是功能方面的决策，对空间的实际应用没有影响。每一个立方体块都通过连续一致的边界来界定，边界与户外的一处具体的景致形成联系；所有独立空间的布局组合在一起，各个空间具有不一样的环境背景，既朝向风景又避免受到相邻景色的影响。从住宅的内部看，其体验是连续的：因为将不同的视野融入多种空间，从住宅的任何角度来看，人们都可以感受到其与环境密切相关。

Sunflower House

The Sunflower house sits on a privileged condition of limit; in the border within the water of the Mediterranean sea and the hard rock of the Costa Brava, between the wild nature of Cabo de Creus and the urban settlement of El Port de la Selva, a small fisherman village is in the border of France and Spain. A place where the Pyrenees get into the water, generating an exceptional wildlife richness, both in the coast and in the water. The house wants to identify each of the particularities of this magnificent landscape; with its geometry, the house frames a multiplicity of different and specific views, and builds up content spaces that inhabit great big framed views.

Mel and Geoff wanted a house in front of the Mediterranean sea that was fully exposed to the views; but they never imagined that their plot, beyond its closeness to the sea, was tremendously exposed to one of the strongest winds of the peninsula (the Tramuntana, up to 180Km/h at this point), and did almost not get any direct sun radiation. So the project starts from this dichotomy: reinforce the relation to the sea, while finding and attracting the sun into the house.

The frontal view from the site is impressive, from France to the Natural Park of the Cabo de Creus, always with the immensity of the open sea right in front of it, the rocks, and a ever changing sky that changes constantly its texture due to the wind.

项目名称：Sunflower House / 地点：Port de la Selva, Girona, Spain
建筑师：Cadaval & Solà-Morales / 项目团队：Eduardo Cadaval, Clara Solà-Morales
合作者：Moisés Gamus, Joanna Pierchala, Efstathios Kanios
建筑工程：Joaquin Peláez / 结构工程：Manel Fernández, BERNUZ-FERNANDEZ
施工单位：Joaquin Gonzalez Obras y Construcciones
总建筑面积：250m² / 有效楼层面积：500m²
设计时间：2012 / 竣工时间：2014
摄影师：©Sandra Pereznieto (courtesy of the architect)

The project breaks down the panoramic view into the addition of many different conditions; the diverse uses of the house are minced and articulated so that each of them (of small dimensions) is positioned frontally to the diverse landscape conditions previously enumerated. Therefore the project is an addition of small units that each frame a differentiated view, and it is within the transition from one unit to the other that the totality of the panoramic view is comprehended. And it is also in the addition of those units where a major open space is generated, the central space of the house.

The house is also a big solar collector, a mechanism to bring light and heat into the house, like a giant sunflower. The composition of the volumes responds to the generation of a rear patio that enables the sun radiation into the living room, to heat the whole house up. This patio, protected from the Tramuntana through the construction itself and oriented to ensure maximum radiation inside the house, is also an outdoor living area where to stay when Tramuntana is hitting the area. Even more, the sequence of two major glazing enables the view of the sea from this back rear patio, while seating within the rocks and local vegetation.

The segmentation into small units is a programmatic decision that has little impact on the actual experience of the place. Each of this cubes is defined through a solid continuous perimeter that traces a specific relationship with the outdoors; the arrangement of all the individual spaces generates an ensemble that reacts to its non-uniform context, opening to the views but protecting itself from neighbors. From the interior the experience of the house is continuous: from any point of the house one feels closely related to the immediate milieu by incorporating one or other view into the numerous spaces.

场地管理 site regulations

地点 location　　朝向 orientation　　视野 view　　风 wind

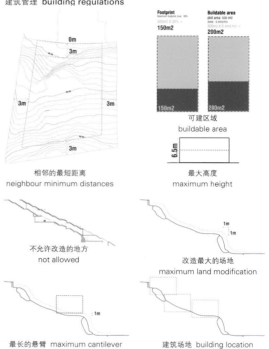

建筑管理 building regulations

可建区域 buildable area

相邻的最短距离 neighbour minimum distances　　最大高度 maximum height

不允许改造的地方 not allowed　　改造最大的场地 maximum land modification

最长的悬臂 maximum cantilever　　建筑场地 building location

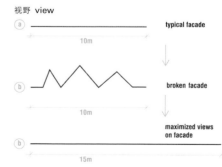

视野 view

typical facade

broken facade

maximized views on facade

可延长的视野 extending views

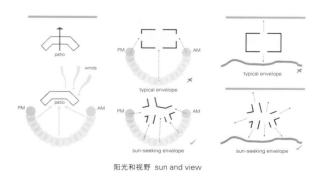

阳光和视野 sun and view

体量研究 volumetric studies

Estudios Volumétricos Modelo_01

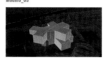 
Estudios Volumétricos Modelo_05

Estudios Volumétricos Modelo_06

Estudios Volumétricos Modelo_09

Estudios Volumétricos Modelo_13

Estudios Volumétricos Modelo_16

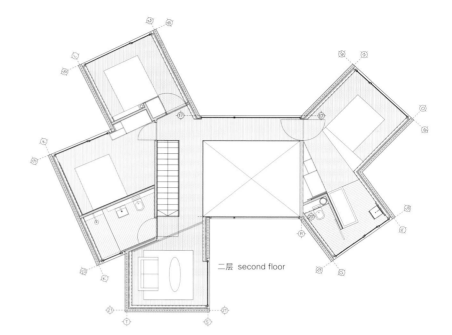

二层 second floor

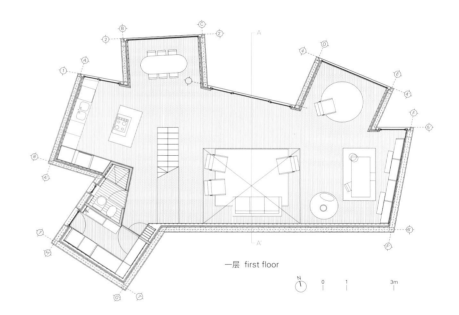

一层 first floor

能量示意图 energetic

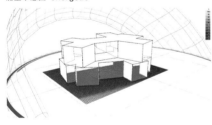

The glass wall towards the south, along with the double height, allows for a flood of natural light throughout the day.

\+ natural light   − natural light

随时间变化的阴影 shadow changes over time

3月21日 March 21　　9月21日 September 21　　12月21日 December 21　　夏至10点 summer solstice @10:00　　12点 @12:00　　18点 @18:00

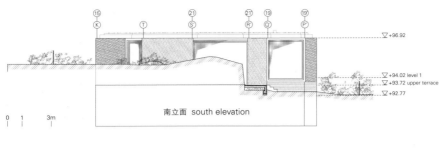

南立面 south elevation

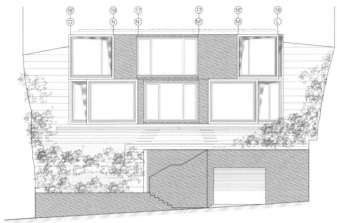

北立面 north elevation

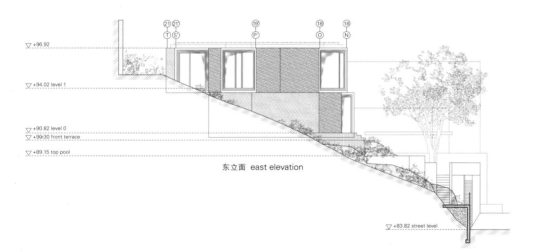

东立面 east elevation

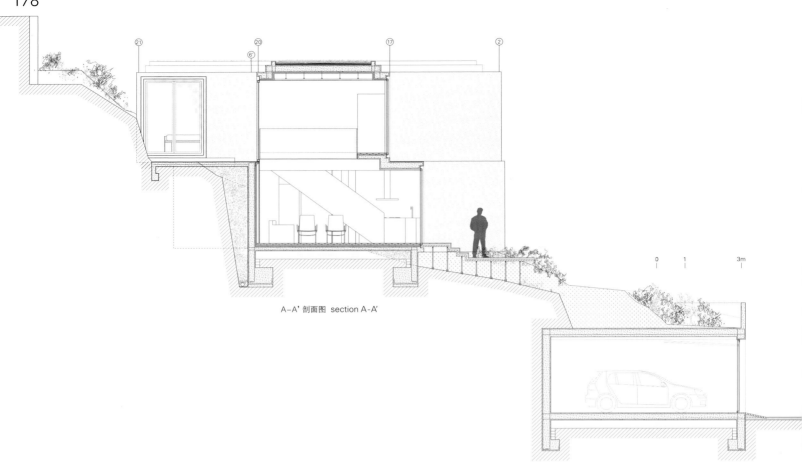

A-A' 剖面图 section A-A'

桧原住宅
SUPPOSE DESIGN OFFICE

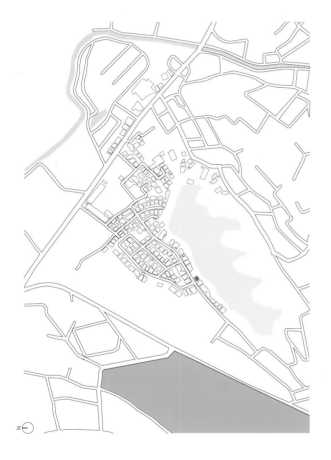

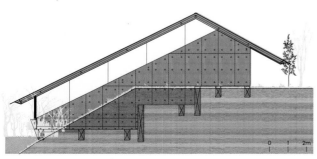

东北立面 north-east elevation

该住宅拥有一个轻便的屋顶。我们希望建造一处在放松休闲或日常生活中都能看到水边的景色的地方。该建造过程很简单。

我们在福冈水库对面的坡地上为一个家庭建造了一个居住的地方。在浓密的花草树木之间，种种生物都向水岸窥探。这个地方使我们感受到了自然的丰富性和舒适。

屋顶下的空间也各式各样，要感受它们之间的差异，取决于你身在何处，空间是封闭还是开放的。

场地的特点，例如斜坡和屋顶，也使居住空间多样化。

进入房间内，首先映入眼帘的是一个半封闭的大厅，但是再往前

走几步远,你就可以看到屋顶下露出的风景优美的湖泊景色。向楼下行走,你便会来到湖边了。景色和氛围也在变化,这里安然宁静,犹如住在田野的小屋内,也似坐在安静的岸边长椅上。

这块空地上的空间是由屋顶,而非墙壁来界定的。屋顶和地面之间的玻璃将此处由户外改造为室内,或者说将室内改造得对环境的变化很敏感。这样,建筑与环境融为一体。

我们意在形成一条分割建筑和环境的界线。我们把建筑视作一个简单的姿态,在空间感知方面具有错综的复杂性。

House in Hibaru

A light roof is put above the site. We would like to make a place that enables the scenery of the waterside to be seen while relaxing, or during our everyday lives. This is done by a simple process.

A living place for a family in the sloping land is made where facing a reservoir in Fukuoka. In between thickly grown trees and flowers, various creatures peek out to the shore. The site makes us feel the richness and comforts of nature.

The roof makes various places. Depending on where you are, spaces close or open, differences can be felt. The character of the site, such as the slope and the roof, makes multiple living places.

Once entering the house, the first thing you see is a semi-closed hall, but a few steps further, you can catch a glimpse of the beautiful lake, framed by the roof. Moving downstairs, the lakeside draws near. The scenery and the atmosphere change. Quietude is felt, like in a small lodge on a field, or on a bench along the calm shore.

The space on the vacant land is defined by the roof, and not by the walls. The glass between the roof and the ground transforms the place from outside to inside, or into an inside that is sensitive to the changes of the environment. Then, architecture assimilates the environment.

The boundary line that divides architecture and its environment becomes the main interest. We try to think the architecture as a simple gesture that generates a complexity in terms of spatial perception.

项目名称：House in Hibaru
地点：Fukuoka, Japan
建筑师：SUPPOSE DESIGN OFFICE
用地面积：259.85m²
总建筑面积：104.94m²
有效楼层面积：111.85m²
结构：steel and RC
设计时间：2011.8—2014.3
施工时间：2014.4—2014.10
摄影师：©Yosuke Harigane (courtesy of the architect)

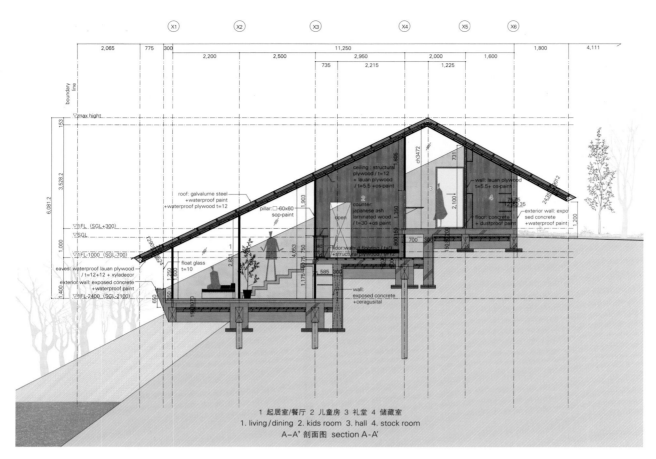

1 起居室/餐厅 2 儿童房 3 礼堂 4 储藏室
1. living/dining 2. kids room 3. hall 4. stock room
A-A' 剖面图 section A-A'

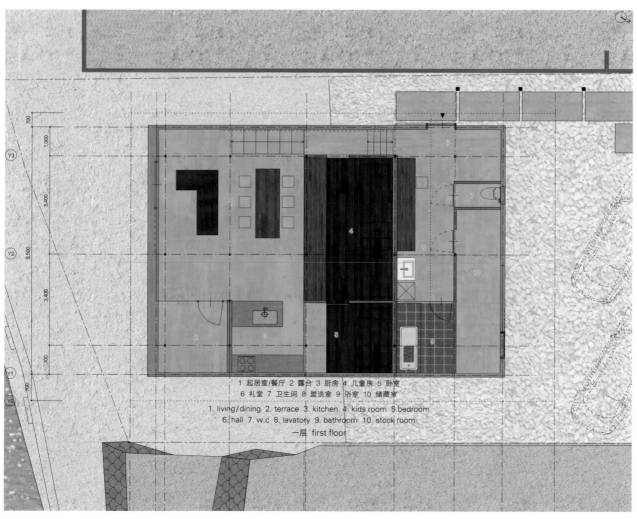

1 起居室/餐厅 2 露台 3 厨房 4 儿童房 5 卧室
6 礼堂 7 卫生间 8 盥洗室 9 浴室 10 储藏室
1. living/dining 2. terrace 3. kitchen 4. kids room 5. bedroom
6. hall 7. w.c 8. lavatory 9. bathroom 10. stock room
一层 first floor

>>68
Riegler Riewe Architekten
Was founded in 1987 in Graz, Austria by Roger Riewe and Florian Riegler, members and registered architects of the Chamber of Architects in Austria, Germany, Poland and Croatia.
Roger Riewe was born in Bielefeld, Germany and studied architecture at RWTH Aachen University. Is currently a professor for Architecture Technology at the Faculty of Architecture at Graz University of Technology.
Florian Riegler was born in Mönichwald, Austria and graduated from the Technical University Graz. Is now teaching design and construction at the University of the Arts(UdK), Germany.

>>170
Cadaval & Solà-Morales
Was founded in New York in 2003 and moved to both Barcelona and Mexico City in 2005.
Eduardo Cadaval holds a BA from the National University of Mexico(2000) and a master's degree from Harvard University(2003). Is an associate professor at ETSAB and a visiting professor at the University of Pennsylvania and the University of Calgary. Clara Solà-Morales also holds a BA from ETSAB(2000) and a master's degree from Harvard University(2003). Is a professor and head of graduate studies at the Barcelona Institute of architecture, visiting professor at the University of Calgary and associate professor at Tarragona's School of Architecture.

>>102
Alberto Campo Baeza
Was born in Valladolid, Spain and grew up in Cádiz. Graduated from ETSAM in 1971. Has been a tenured professor for more than 25 years of ETSAM. Has taught at the ETH in Zurich and the EPFL in Lausanne as well as the University of Pennsylvania in Philadelphia, the Kansas State University, and the CUA University in Washington.
And he has given many lectures all over the world, and has received many awards. Has completed a number of significant projects. His works have been published in major architectural magazines and exhibited in major cities.

>>10
51N4E
Is a Brussels-based international practice founded in 1998. Is led by two partners; Johan Anrys, Freek Persyn. And 28 people are working at the office. Johan Anrys and Freek Persyn were both born in 1974 in Belgium and studied architecture in Brussels and Dublin, graduating in 1997. Aspires to contribute, through means of design, to social and urban transformation. Concerns itself with matters of architectural design, concept development and strategic spatial transformations.

>>82
Lacaton & Vassal
Anne Lacaton[left] and Jean Phillippe Vassal[right] founded the office Lacaton & Vassal in 1989, based in Paris.
Anne Lacaton, graduated from the School of Architecture of Bordeaux in 1980 and received a diploma in Urban Planning from the University of Bordeaux in 1984. She has been teaching at the University of Madrid since 2007.
Jean Philippe Vassal was born in Casablanca, Morocco and graduated from the School of Architecture of Bordeaux in 1980. He has been teaching at UdK Berlin since 2012.

>>152
Gonzalo Mardones Viviani
Graduated from the Catholic University of Chile in 1980 with the Maximum Honors. In 1979, he founded Mardones Arquitectos y Asociados with Carlos Mardones Santistevan and created Gonzalo Mardones Viviani Arquitectos in 1999. Directed degree projects as professor in architectural design at the same university and in several other universities. He has been working as guest professor and lecturer until today.

>>160
Fougeron Architecture
Is a nationally recognized design firm whose work exhibits a strong commitment to clarity of thought, design integrity and quality of architectural detail. The firm's decidedly modernist attitude is the result of founder Anne Fougeron's vision to create a practice dedicated to finding the perfect alignment between architectural idea and built form.
Anne Fougeron received a BA from the Wellesley College, Massachusetts in 1976 and M.Arch from the University of California, Berkeley in 1980. Before establishment of Fougeron Architecture in 1986, she has worked at San Jose Redevelopment Agency and Daniel Solomon and Associates. She

>>116
João Paulo Loureiro
Was born in City of Porto, Portugal in 1970 and studied at the School of Architecture of the University of Porto. Established his own office at Matosinhos called "JPLoureiro, Architect, Inc" in 2006. His work was presented at the national exhibition Habitar Portugal 2006/08. Is a member of the board of directors of the Professional Architects' Association in Northern Portugal between 2005 and 2010. In 2013, he was invited as Assistant Professor at the Oporto School of Architecture(FAUP).

>>182
SUPPOSE DESIGN OFFICE
Was established by Makoto Tanijiri in Hiroshima, 2000. Is now led by Makoto Tanijiri and Ai Yoshida. Their work covers a broad range of areas including designing houses, business spaces, frameworks, landscapes, products, and art installations. Makoto Tanijiri defines his work as a chance to realize fresh ideas about buildings and relationships of all interactive elements. They received numerous prizes including JCD Design Award, Good Design Award, DFA Award and so on.

>>132
Sebastián Irarrázaval Arquitectos
Sebastián Irarrázaval was born in 1967. He studied at the Catholic University of Chile and the Architectural Association in London. In 1993, he set up his own practice in Santiago. Received the AOA(Architecture Offices Association) award for the most outstanding young architects and he was also awarded in the XVI Architecture Biennial in Chile. He has been teaching design studio at the Catholic University of Chile since 1994. His projects have been published worldwide in specialized magazines, among them: Casabella, Arquitectura Viva, C3, ARQ and A+U. Practice work has also being exhibited locally and abroad. Recent exhibitions include XV Chilean Architectural Biennial, Venice Biennale, Shenzhen and Hong Kong bi-city Biennial. This year he was awarded at the Wave International Workshop held at IUAV in Venice.

>>36
O-office Architects
Was established in 2007 by He Jianxiang[right] and Jiang Ying[left] in Guangzhou. They approach architecture across various fields like technology, arts and media on the overwhelming background of "production" in contemporary China.
He Jianxiang worked for VK Group, Belgium, as a project architect until 2003 and established O-office in 2004. He received a M. Arch.(2009) and M. S. in Arch in K.U.(2010) Leuven, Belgium. Is currently teaching at the Guangzhou Academy of Fine Art.
Jiang Ying graduated from National School of Architecture in Versailles and worked for AREP in Paris as a project architect after graduation. And he became a D.P.L.G(Diplôme par le Gouvernement) in 2004.

Jack Self
Is an architect and writer based in London. Is a director of the Real Foundation, Contributing Editor for the Architectural Review and also an author of Real Estates: Life Without Debt (Bedford Press, 2014).

Nelson Mota
Graduated from the University of Coimbra, where he lectured from 2004 to 2009. Currently he is an Assistant Professor at the TU Delft, in the Netherlands, where he concluded his Ph.D in 2014 with the title "An Archaeology of the Ordinary. Rethinking the Architecture of Dwelling from CIAM to Siza". He was the recipient of the Tavora Prize(2006) and authored the book A Arquitectura do Quotidiano(2010). He is a member of the editorial board of the academic journal Footprint, and a founding partner of Comoco architects.

>>20
João Mendes Ribeiro Arquitecto
João Mendes Ribeiro was born in Coimbra, Portugal, 1960. Graduated from the Faculty of Architecture of the University of Porto, where he taught between 1989 and 1991. He took his Ph.D. in Architecture from the University of Coimbra, in 2009, in the field of Theory and History. His work has been widely published, both in national and international publications. He took part in a large number of national and international exhibitions, including the 9th and 10th Venice Biennale, the 11th Prague Quadrennial of Performance Design and Space and the 7th São Paulo Architecture Biennial.

C3, Issue 2015.11
All Rights Reserved. Authorized translation from the Korean-English language edition published by C3 Publishing Co., Seoul.

© 2016大连理工大学出版社
著作权合同登记06-2016年第98号
版权所有·侵权必究

图书在版编目(CIP)数据

家居生态：汉英对照 / 韩国C3出版公社编 ；杜丹等译. — 大连：大连理工大学出版社，2016.7

(C3建筑立场系列丛书)
书名原文：C3：Ecologies of Domesticity
ISBN 978-7-5685-0455-3

Ⅰ．①家… Ⅱ．①韩… ②杜… Ⅲ．①住宅－景观设计－汉、英 Ⅳ．①TU241
中国版本图书馆CIP数据核字(2016)第168134号

出版发行：大连理工大学出版社
　　　　　（地址：大连市软件园路80号　邮编：116023）
印　　刷：上海锦良印刷厂
幅面尺寸：225mm×300mm
印　　张：12.25
出版时间：2016年7月第1版
印刷时间：2016年7月第1次印刷
出 版 人：金英伟
统　　筹：房　磊
责任编辑：张昕焱
封面设计：王志峰
责任校对：高　文
书　　号：978-7-5685-0455-3
定　　价：228.00元

发　　行：0411-84708842
传　　真：0411-84701466
E-mail：12282980@qq.com
URL：http://www.dutp.cn